Greening the Alliance

Greening the Alliance

The Diplomacy of NATO's Science and Environmental Initiatives

SIMONE TURCHETTI

The University of Chicago Press   Chicago and London

The University of Chicago Press, Chicago 60637
The University of Chicago Press, Ltd., London
© 2019 by The University of Chicago
Published 2019
Printed in the United States of America

27 26 25 24 23 22 21 20 19 18   1 2 3 4 5

ISBN-13: 978-0-226-59565-8 (cloth)
ISBN-13: 978-0-226-59579-5 (paper)
ISBN-13: 978-0-226-59582-5 (e-book)
DOI: https://doi.org/10.7208/chicago/9780226595825.001.0001

Library of Congress Cataloging-in-Publication Data

Names: Turchetti, Simone, author.
Title: Greening the alliance: the diplomacy of NATO's science and
    environmental initiatives / Simone Turchetti.
Description: Chicago; London: The University of Chicago Press, 2019. |
    Includes bibliographical references and index.
Identifiers: LCCN 2018029686 | ISBN 9780226595658 (cloth: alk. paper) |
    ISBN 9780226595795 (pbk.: alk. paper) | ISBN 9780226595825 (e-book)
Subjects: LCSH: North Atlantic Treaty Organization—History. |
    Environmental Sciences—History—20th century. | Environmental
    protection—Research—History. | Environmental sciences—Political
    aspects. | Research—History—20th century. | Research—Political aspects.
Classification: LCC GE50 .T86 2018 | DDC 304.209182/1—dc23
LC record available at https://lccn.loc.gov/2018029686

♾ This paper meets the requirements of ANSI/NISO Z39.48–1992
(Permanence of Paper).

We live in an atomic age, Mr. Wormold. Push a button—piff bang—where are we? Another Scotch, please.

GRAHAM GREENE, *OUR MAN IN HAVANA*

Contents

Abbreviations

ACE	Allied Command Europe (NATO)
AEC	Atomic Energy Commission (US)
AFCRC	Air Force Cambridge Research Center (US)
AFCRL	Air Force Cambridge Research Laboratories (USAF, ex-AFCRC)
AGARD	Advisory Group for Aeronautical Research and Development (NATO)
ASI	Advanced Study Institute (NATO)
ASTIA	Armed Forces Technical Information Agency (US)
AWACS	Air-Borne Early Warning and Control System
CCMS	Committee on the Challenges of Modern Society (NATO)
CENTO	Central Treaty Organization
CEQ	Council for Environmental Quality (US)
CERN	European Organization for Nuclear Research
CIA	Central Intelligence Agency (US)
CNET	Centre National d'Études des Télécommunications (France)
COSPAR	Committee on Space Research (ICSU)
DoD	Department of Defense (US)
DRD	Defence Research Directors (NATO)
DRDC	Committee of Defence Research Directors (NATO)
DRG	Defence Research Group (NATO)
DSE	Department of Science and Education (UK)
EARSeL	European Association of Remote Sensing Laboratories
EOARDC	European Office of Air Research and Development Command (USAF)
FLIP	Floating Instrument Platform
GARP	Global Atmospheric Research Programme (ICSU/WMO)
IAEA	International Atomic Energy Agency
ICBMs	Intercontinental Ballistic Missiles
ICES	International Council for the Exploration of the Sea

ICSU	International Council of Scientific Unions
IGY	International Geophysical Year (ICSU)
IIST	International Institute of Science and Technology (NATO)
IMCO	Intergovernmental Maritime Consultative Organization
IOS	Institute of Oceanographic Sciences (ex-NIO)
IPCC	Intergovernmental Panel on Climate Change (UNEP/WMO)
JASIN	Joint Air Sea Interaction Experiment (GARP)
JCS	Joint Chiefs of Staff (US)
JIGSAW	Joint Inter Service Group for the Study of All-out Warfare (UK)
JONSWAP	Joint North Sea Wave Project
LDC	Less Developed (NATO) Countries
MAFF	Ministry of Agriculture, Fisheries and Food (UK)
MARSEN	Maritime Remote Sensing
MRBMs	Medium-Range Ballistic Missiles
MWDP	Mutual Weapons Development Program (US)
NAC	North Atlantic Council (NATO)
NACA	National Advisory Committee on Aeronautics (US)
NADGE	NATO Air Defense Ground Environment
NAS	National Academy of Sciences (US)
NASA	National Aeronautics and Space Administration (US)
NATIS	NATO Information Service
NATO	North Atlantic Treaty Organization
NDRE	Norwegian Defence Research Establishment
NIO	National Institute of Oceanography (UK)
NORSEX	Norwegian Remote Sensing Programme
NSF	National Science Foundation (US)
OECD	Organization for Economic Co-operation and Development (ex OEEC)
ONR	Office of Naval Research (US)
OPAQUE	Optical Atmospheric Quantities in Europe
ORBIS	Orbiting Radio Beacon Ionospheric Satellite (AFCRL)
RAF	Royal Air Force (UK)
SAC	Strategic Air Command (US)
SACEUR	Supreme Allied Commander for Europe (NATO)
SACLANT	Supreme Allied Commander for the Atlantic (NATO)
SACLANTCEN	SACLANT Anti-Submarine Warfare (ASW) Research Centre
SADTC	SHAPE Air Defense Technical Centre (NATO)
SCEP	Study of Critical Environmental Problems
SCOR	Scientific Committee on Oceanographic Research (ICSU)
SEATO	South-East Asian Treaty Organization
SHAPE	Supreme Headquarters of Allied Powers in Europe (NATO)
SLBMs	Submarine-Launched Ballistic Missiles
SPS	Science for Peace and Security (NATO)
TNO-RVO	Dutch Defense Research Establishment

UN	United Nations
UNEP	UN Environmental Programme
UNESCO	UN Educational, Scientific and Cultural Organization
UNFCCC	UN Framework Convention on Climate Change
USAF	US Air Force
USIA	US Information Agency
VKC	Von Kármán Committee
WCRP	World Climate Research Program (WMO-ICSU)
WEU	Western European Union
WHOI	Woods Hole Oceanographic Institution (US)
WMO	World Meteorological Organization
WSEG	Weapon Systems Evaluation Group (US DoD)

Archives Consulted
(and Abbreviations Used
to Identify Them)

DEA George Deacon Papers, National Oceanography Library, Southampton, UK

CAG Papers of the Consiglio Nazionale delle Ricerche—CNR, Vincenzo Caglioti Directorship (1965–1972), Archivio Centrale dello Stato, Rome, Italy

ING Istituto Nazionale di Geofisica Archive, Istituto Nazionale di Geofisica e Vulcanologia, Rome, Italy

JBA Jodrell Bank Archive, University of Manchester Library, Manchester, UK

KOE Joseph Koepfli Papers, California Institute of Technology Archive, Pasadena, California

MOS Håkon Mosby Papers, University of Bergen, Bergen, Norway

NARA National Archives and Records Administration, College Park, MD

NATO NATO Archives, NATO Headquarters, Brussels, Belgium

PAT Clair Cameron Patterson Papers, California Institute of Technology Archive, Pasadena, California

RABI Isidor Isaac Rabi Papers, US Library of Congress, Washington, DC

ROB Howard Percy Robertson Papers, California Institute of Technology Archive, Pasadena, California

TNA The National Archives, Kew Gardens, London, UK

TRA Russell Train Papers, US Library of Congress, Washington, DC

TVK Theodore von Kármán Papers, California Institute of Technology Archive, Pasadena, California

ZUC Solly Zuckerman Papers, University of East Anglia, Norwich, UK

Introduction: NATO's Imperatives

On November 26, 2011, in Durban, South Africa, one of the most decisive meetings in the recent history of international relations opened. Delegates at the Seventeenth Conference of Parties of the UN Framework Convention on Climate Change (UNFCCC) were about to enter negotiations on carbon emission cuts in order to confront the threat of global warming. But on that same day, another event took away the media's attention: a complex surveillance operation near a military checkpoint fifteen thousand kilometers away from Durban. In a scarcely populated district of the Federally Administered Tribal Area at the Pakistan border with Afghanistan known as Salala, a NATO unit identified a Taliban convoy crossing the area and attacked it. The soldiers then realized to their horror that the party did not carry Taliban fighters but consisted of regular Pakistani soldiers. Twenty-two Pakistanis died, ratcheting up tensions between their administration and NATO allies.

To the reader, the contemporaneous UNFCCC meeting and the Salala checkpoint incident may appear completely unconnected, but, in fact, they were crucially linked. International negotiations on climate change have followed the gathering of compelling scientific evidence through the use of sophisticated technologies showing the warming of our planet. Similar instruments have been used for several decades to improve surveillance operations such as the ones that anticipated the Salala attack. Electronic signals have worked as diagnostic tools to scan landscapes and airspaces, sometimes to detect the presence of enemies, other times to assess environmental threats.

In our rapidly changing world, the daily reliance of government agencies upon invisible yet ubiquitous electronic systems for remotely observing and analyzing the environment has grown considerably. As electromagnetic pulses travel back and forth between satellites and earth-based transmitters and receivers, the amount of data on the status of our planet massively increases and so, too, does the intelligence on the humans inhabiting the earth. Both become decisive in the context of international affairs, informing key decisions on local and global issues. The experts involved in this kind of remote observation of both humans and environments have coined the term *remote sensing* to describe it—the use of technological devices as electronic eyes and ears to "sense," from distance, the world around us. Some have even been explicit about their underlying ambitions, dubbing such practices "remote surveillance."[1]

Although the number of national and international organizations involved in amassing new data and knowledge through remote sensing is significant, only one to my knowledge has, throughout its history, invested substantially and continuously in improving research methods *both* for monitoring environmental change *and* for defense purposes. That organization is the North Atlantic Treaty Organization (NATO). Established in 1949 to counter the perceived threat of a Soviet invasion, for the last sixty years this defense alliance has been a prominent patron for scientific research and environmental studies. Starting in 1958, the alliance's science program propelled the funding of scholarships, scientific meetings, and collaborative research exercises. Ten years later, NATO entered a new phase in its history typified by a program of research and actions addressing environmental problems; something that entailed the application of remote-sensing methods to the charting of changes in the environment.

Based on a copious amount of untapped archival records from several repositories, including NATO's archive, *Greening the Alliance* is the first comprehensive account exploring the history of these programs from their inception to the present. The volume shows why the natural environment featured as a key research focus for the alliance during the past sixty years and how the organization of NATO-wide research collaborations shaped its trajectory as a multilateral agency.

For quite some time, the only sources documenting the role of NATO as patron of science and environmental research have been specialized internal reports written by former officials.[2] More recently, some scholars have started to pay greater attention to NATO's patronage. John Krige has convincingly documented its origins and argued that the alliance's investment in science aimed to generate consensus on a North American model of knowledge production.[3] Jacob Hamblin has discussed the circumstances of NATO's later environmental research in two works: *Arming Mother Nature* discusses its environmental warfare plans, and a recent article investigates its promotion of actions against environmental degradation.[4] Hamblin's research has provided valuable insights. But it has left many open questions, especially as it does not fully explain the coexistence of competing ambitions: how could NATO's officials configure the environment's potential in changing warfare while, at the same time, investing in its protection?

This volume answers this question and provides a novel understanding of the motive behind NATO's investment. The examination of the alliance's sponsorship patterns has now revealed an inclination and a strategic core of studies that was initially not open to public scrutiny. By looking at the distribution of research funds and the setting up of research groups devoted to specific tasks, the staple of its sixty-year-long investment is finally in our sight and is herewith revealed.

NATO's defense and surveillance priorities in the increasingly globalized war theater that emerged from the end of World War II drove the sponsorship of studies on the physical characteristics of natural environments. Having to assess the presence and capability of enemy forces, especially after the launch of the Soviet satellite Sputnik, the alliance's commanders instructed the setting up of powerful surveillance systems on land, in air, and underwater. As these systems were being developed, the experts that NATO appointed understood that their performance depended on an environmental medium. Cloud cover blocked radar scanning the airspace to detect enemy aircraft; changes in temperature deflected sound pulses emitted by sonar seeking submarines hiding in sea waters. These operational imperatives informed NATO's research agenda, also paving the way to prioritizing the funding of disciplines like oceanography and meteorology.[5]

NATO's distinctive patronage thus aligned to its role in the Cold War. Its leaders embraced an approach to defense that emphasized the virtues of rational planning and the adoption of modern science and technology. Their Cold War mind-set has been convincingly described in a number of recent works discussing the trajectory of fields as diverse as operations research, psychology, and economics.[6] NATO's operational approach also redefined the environment as something that—not differently from any other weapon system—could be divided into distinct processes. The scientists responsible for NATO projects also concluded that terrestrial phenomena could be fully understood and, under certain circumstances, even manipulated. Control of nature as manifested through environmental warfare was thus the starkest example of how NATO's defense imperatives shaped its sponsorship.

So in the realms of both surveillance and environmental management, the defense alliance funded studies emphasizing the merits of remote sensing and the virtues of prediction of natural events in light of improved monitoring techniques. NATO-sponsored scientists pioneered "at distance," "impersonal" analyses of physical environments (typically associated with the rise of geophysics as a scientific discipline), defining nature as a synoptic subject of inquiry to be studied through remote-sensing methods.[7] This approach defined their work as antithetical to that of other experts emphasizing instead the distinctiveness of local environments, their uniqueness, and the need for experimenters of direct contact with nature.

It would be wrong, however, to assume that NATO's science and environmental programs were shaped by the alliance's military commanders alone (or the scientists they funded). Their objectives were set out in tense negotiations between member states' representatives. This changes, in my view, our understanding of how patronage informed the process of scientific change, especially during the Cold War. Paul Forman's archetypal study on the sponsorship of quantum electronics in the United States was the first to claim that national security imperatives significantly "warped" the scientists' research agenda and defined new priorities aligned to defense needs.[8] Daniel Kevles and others objected, though, claiming that scientists were not just pawns; they were actively involved in shaping these programs.[9] These interpretations have informed other studies, including those devoted to the history of the earth and environmental sciences.[10]

This book tackles the issues that Forman and Kevles have previously grappled with but emphasizes the importance of the negotiation phase without assuming that it was always the sponsors or the grant recipients who dictated what to study. NATO's configuration as a multilateral organization meant that priorities were set on the basis of what pieces of scientific knowledge the allies intended to share; what they viewed as decisive to national research; and, especially, what else they were prepared to embark upon in order to please or appease other allies. I have referred elsewhere to this give-and-take in terms of "trading exercises," which, by the time NATO launched a science program, already typified bilateral negotiations between its member states.[11]

So it was mainly the effort to overcome disagreements that gave leeway to NATO's patronage. In particular, scientists representing the US delegation at NATO were particularly eager to embark on a surveillance-oriented research program, but their Western European colleagues challenged these stances on what research the alliance should pursue. Competing views existed, more generally, on the structure of its "military-scientific complex," and although US representatives took for granted that a defense research focus would propel Western science, their colleagues in Europe indicated this forecast as naive.[12] And because NATO's decisions about the science program affected the advancement of Western science as much as the alliance's cohesiveness, this book also shows that the consensual definition of NATO-sponsored studies was equally decisive in its history as a multilateral organization.

Science at NATO: Diplomacy by Other Means?

NATO has been for several decades an important subject of historical research. Established to be "sword and shield" against the Soviet-dominated Eastern Bloc, it considerably expanded after that, eventually becoming foremost a defense alliance in the age of globalization. Several generations of scholars have explored its history and, understandably, focused on its military structures and weapon systems, commands and commanders, state officials and diplomats.[13] Lawrence Kaplan's archetypal and compelling study was the first to comprehensively analyze its defense and political trajectories.[14] Common security, especially in relation to nuclear weapons, has been the focus of several works too.[15]

Few have, however, examined NATO's science and environmental programs with a view to explaining them as critical components of its history. This book reveals these schemes and the negotiations that anticipated their elaboration to have been absolutely decisive in the life of this organization; especially as they aimed to unlock otherwise jammed relations between member states through the offer of scientific collaboration. The promotion of science was, in essence, a key device of NATO diplomacy.

Science diplomacy is a term that has recently acquired currency and identifies, among other things, how the promotion of scientific collaboration between national research groups can be a powerful "soft power" device to strengthen relations between two or more countries.[16] Learned societies such as the American Association for the Advancement of Science (AAAS) in the United States and the Royal Society in the United Kingdom have recently underscored the importance of science in diplomatic work and organized initiatives to further deepen its understanding.[17] During one of these initiatives, one NATO official even framed the alliance as "instrumental in the history of science diplomacy."[18]

While it is up to the reader to judge whether NATO really had such a pivotal role, *Greening the Alliance* invites a reconsideration of the portrayal of science diplomacy as a force in international relations, anchoring this concept more firmly to the actualities of Cold War geopolitics. As Marc Trachtenberg has argued, consensus at NATO often hung on a thread of minor propositions while the allies disagreed on everything else; the Atlantic alliance—Ralph Dietl shows—has been "a troubled partnership" for most of its history.[19] The promotion of scientific collaboration helped officials representing national governments to evade thorny issues marring the cohesion of the alliance in key periods of its history and to establish one domain to build consensus.[20] The promotion of science was, in essence, "back-channel" or "parallel" diplomacy; that is, the construction of an alternative conduit to constructive relations as these appeared to crumble.[21]

By the late 1950s, NATO was much in need of diplomatic back channels especially in light of the difficulties preventing the alliance's integration. French and British officials disputed whether the United States alone should control its nuclear deterrent. Smaller European countries such as Italy and Turkey agreed to host this deterrent notwithstanding French and British anxieties.[22] West Germans agreed with their US colleagues on Germany's disarmament but secretly hoped to share a nuclear deterrent with France. Norway, the Netherlands, and Denmark supported US stances but opposed efforts to locate nuclear weapons on their soil. In this complex diplomatic landscape, solutions were not easy to find. One of the central characters in the book, the Nobel Prize–winning US scientist Isidor Isaac Rabi, at one point solemnly proclaimed to be done with "the whole ball of wax which is NATO."[23]

Indeed the alliance often seemed to be melting in the heat of divergent imperatives, which is the reason scientists like Rabi viewed the promotion of a science program as something that could lessen NATO's political frailty. Furthermore, by the time this promotion started, scientists had just demonstrated distinctive ambassadorial skills. In 1957, especially because of the Suez crisis, NATO's traditional enthusiasm for diplomatic work was at a low ebb. By contrast, one of the most important episodes in the history of international scientific cooperation, the International Geophysical Year (IGY; 1957–58, which also had a research focus on the natural environment), had attracted growing media attention. This was partly because it had revealed the scientists' potential to innovate international relations through the coordination of novel research. Even Eastern Bloc scientists were persuaded to contribute.

The completion of the Soviets' main contribution to this event, the Soviet satellite Sputnik, made US officials more anxious about the IGY. But they also realized that scientists, internationally, now seemed to speak a more persuasive language than traditional diplomats, partly because of science's propagandized universality in a time of growing global inter-dependence of nation-states.[24] And the growing number of multilateral agencies pursuing scientific research increased NATO's eagerness to create its program and compete with those of other international organizations. The North Atlantic Council (NAC; NATO's chief decision-making struc-ture) now instigated the appointment of scientists as negotiators to iden-tify suitable research themes for sponsorship using both scientific and political criteria as discriminant factors. The NATO Science Committee, composed of scientists as national delegates, was thus established to pro-vide advice on a science program. Rabi and other experienced scientists featured in this volume (such as the US expert Howard P. Robertson, the Briton Solly Zuckerman, the Frenchman Louis Néel, the Norwegian Gun-nar Randers, the Italian Amedeo Giacomini, and the Turk Nimet Özdas) were called in to find proposals that could bring otherwise alienated allies together, thus enmeshing—as "science diplomats"—scientific and diplo-macy roles.[25]

So aside from capturing one key aspect of NATO's history as an organi-zation, this book also suggests revisiting the idea that the US was its hege-monic power.[26] True that the proposals of US delegates to stir its research efforts often prevailed, but they were put forward in an attempt to dodge more compelling issues dividing the allies and could succeed only when the delegates found ways to shape new (and often fleeting) alliances on specific items with colleagues from other countries. Leadership thus rested with these transitory partnerships rather than with one nation alone.

One important crisis that hit the alliance in the second half of the 1960s demonstrates this further. By then US stances on integration and consensus building through the promotion of science found stronger re-sistance in Western Europe. British and French delegates articulated their opposition in terms of reluctance to further fund the NATO science pro-gram. It was this crisis that reconfigured its science diplomacy as "envi-ronmental."

Greening: The Transition to Environmental Diplomacy

At the height of the Cold War, NATO decided to invest substantially in research on environmental protection and conservation as if the emerald green of environmentalism could, metaphorically, go hand in hand with the camouflaging khaki of its defense organizations. This entangling of environmental and defense ambitions may strike the reader as incongruous, as scholars interested in twentieth century history and environmental history have typically associated the origins of environmentalism with grassroots organizations like Greenpeace and other antinuclear groups known for their nonviolent (rather than peacekeeping) credentials.[27] Actually, the protection, conservation, and management of natural resources have often been considered contiguous with the activities of those groups that, especially in Western Europe, have criticized NATO and its role in Cold War affairs.

Yet NATO's environmental ambitions represented, as much as the promotion of science, a device for improving diplomatic relations. J. Brooks Flippen, Jacob Hamblin, and Stephen Macekura, among others, have used the notion of environmental diplomacy to underline how, from the late 1960s, nature conservation and environmental protection played a more prominent role in international relations. No longer solely the domain of grassroots groups outside mainstream politics, environmentalism took center stage in the international political arena because of the direct involvement of national governments.[28] The environment had featured in foreign affairs before, but the shift occurring at the end of the decade was significant nonetheless.[29] International discussions on environmental protection renewed collaborations and tensions already existing at a national level between state administrators, environmentalists, and scientists.[30]

The search for cohesiveness that had already typified the emergence of a science program also explains the alliance's "environmental turn." In the mid-1960s, US officials were busy looking for ways to innovate the political debate in light of the crisis of the Science Committee and NATO more generally. By that time, the representatives of NATO's less developed countries had been particularly vocal, threatening to withdraw support from the Science Committee.[31] And the French, dissatisfied with US leadership on military coordination, were about to abandon the alliance's command structure.

Officials at the US State Department now thought of new schemes to constructively evade NATO's problems; one such plan consisted of forging consensus on the need for the alliance to promote environmental actions. The North Atlantic Council eventually instructed setting up the new Committee on the Challenges of Modern Society (CCMS), while the Science Committee was reformed so as to prioritize environmental research. The focus of NATO's parallel diplomacy now shifted from the promotion of science alone to environmental research and management, which coincided with the appearance on the NATO scene of the celebrated US conservationist Russell Train also to play (as Rabi and other scientists had done that far) a diplomatic role.

The alliance's "greening" thus consisted of, first and foremost, taking to completion a political restoration project that at its core presented diplomacy rather than environmental objectives.[32] In so doing, NATO's patronage of environmental research and actions forged a distinctive, idiosyncratic even, type of environmentalism. Frank Zelko has aptly defined Greenpeace's environmentalism as "countercultural" to underscore the fact that environmental campaigning led its activists to challenge dominant views on science and technology as unequivocally benign forces in the struggle against environmental degradation.[33] Conversely, NATO officials endorsed environmentalism that married scientific rationalism. As Andrew Jamison and Stephen Bocking have pointed out, in the late 1960s the interaction between scientists and decision makers was decisive in reframing environmental debates reinforcing "assumptions about society, the world, and the possibility of controlling and predicting events and developments in the world."[34] NATO's entrance in the environmental arena paved the way to reinforcing some of these assumptions from an "establishment" perspective as support to scientific institutions and researchers within the context of the advanced capitalist societies distinctive of NATO's member states would have unquestionably led to successfully tackling environmental change. NATO's sponsorship thus worked as a countervailing force to the grassroots movements like Greenpeace and their radical stances.

But the underlying diplomacy ambitions of NATO's environmentalism configured the limits of its actions. Propositions compromising diplomatic efforts were relinquished or rendered ineffective, whereas those generating consensus within the alliance were prioritized regardless of whether they were timely, sound, or feasible. On the whole, the alliance continued to harbor serious inconsistencies. During the 1970s and 1980s, the enmeshing of environmental politics and international diplomacy did not prevent national interests from exerting a strong influence: stricter environmental regulations failed to receive immediate approval and were endorsed only after being technically watered down. NATO's trading exercises in environmental diplomacy did not, therefore, always produce results aligning to environmental protection.

The political qualities of NATO's environmentalism continued to be reaffirmed by its promoters, however, even as the Cold War grew tenser. New projects even coupled innovative research on environmental threats with traditional surveillance-oriented work, while NATO's establishment environmentalism went hand in hand with the nuclear rearmament of Western Europe and the promotion of nuclear energy. With the end of the Cold War, as the final chapters of the book show, NATO officials realized even better the importance of keeping scientific and environmental research going because of the need to foster friendly relations with former Eastern Bloc countries. It is only fairly recently that NATO's programs have gone through a comprehensive review. But this has happened in new geopolitical circumstances and at a point in its history when the alliance's patronage has changed considerably.

Greening the Alliance reflects on the legacy of this sponsorship, another theme that has kept scholars busy in light of our current anxiety with global environmental challenges, and especially climate change.[35] If it was the Cold War that originally shaped studies of the natural environment, then what does this mean for the current configuration of environmental studies and policies today? What have we inherited and what have we abandoned? Naomi Oreskes has defined the shift in research focus from traditional Cold War tasks to the analysis of environmental change as a "new mission."[36] She has warned that reading continuities between Cold War science and present research efforts as causal relations can be misleading, but she has also highlighted that in the last thirty years the growing attention to environmental degradation, leading up to the global warming debate, has offered an opportunity for environmental scientists to redefine their research interests and find new sponsors.

NATO's example shows interesting tensions as early efforts to promote climate change research were met with resistance. Furthermore, when both science and environmental diplomacy proved less successful, the alliance started downsizing its patronage. The nearly contemporaneous Salala incident and UNFCCC meeting in Durban are not only crucially linked, but they actually epitomize the tensions within NATO's historical engagement with scientific research and environmental action. They display the clash between the two shades of green that one could say have colored its programs. It is this clash that this book portrays.

Chapter Overview

This historical study explores the history of NATO's science and environmental programs from the 1950s to the turn of the century. Chapter 1 looks at NATO's circumstances before the setting up of a Science Committee; a period typified by a limited investment in novel research in the context of the alliance's defense research agencies. By the end of the 1950s, the resonance of the IGY (and Sputnik) and the problems of integration of the alliance convinced its administrators to invest more in science and technology through the creation of a Science Committee. Chapter 2 shows how this organization accommodated NATO's defense and surveillance ambitions in its sponsorship as a consequence of the creation of the Science Committee's subgroups on oceanography, meteorology, and radio meteorology that were devoted to the analysis of the weather, the atmosphere, and the seas surrounding Europe in connection with NATO's operational requirements.

By the mid-1960s, prominent NATO scientists manifested reservations about the Science Committee program and what the scientific projects it had sponsored had (or had not) so far accomplished. Chapter 3 illustrates how this criticism paved the way to the Science Committee's crisis, which was heightened by the divergence of opinion on the alliance more generally by its more prominent member states. The possibility of engaging in an environmental warfare program further escalated the controversy on the committee's achievements.

Chapter 4 displays how a few officials in the US State Department conceived solutions to the committee's (and NATO's) crisis and highlights how a turn toward research and actions against environmental degradation was envisaged as a way to take the committee out of the quagmire while evading other issues dividing the allies. This stance coalesced in the creation of the CCMS, and from 1970, NATO's science program was reconfigured as consisting primarily of environmental research. Chapter 5 looks at how the CCMS activities helped redefine the Science Committee structure during the 1970s, through the dissolution of old defense-oriented subgroups and the setting up of new ones devoted to environmental research.

We learn in chapter 6 that NATO research groups interested in environmental and traditional defense-oriented research ended up having interesting exchanges on specific issues enmeshing NATO's "green" and "khaki" dimensions. The chapter examines therefore the entanglement between surveillance-driven and environment-related research, as well as, more generally, the development of the NATO science and environmental programs during the 1980s.

Chapter 7 examines the trajectory of NATO science at the dawn of the new century focusing on two overlapping issues: the search for ways to further integrate and strengthen the alliance through its new "Science for Stability" program and the ambivalence of some national representatives toward environmental research and programs. The final chapter (chapter

8) looks into the current state and future of NATO as a patron for science and environmental research. It also draws some conclusions about the distinctiveness of its sponsorship in the first fifty years of its existence.

A Note on Sources and Acknowledgments

Greening the Alliance results from several years of research, primarily on recently disclosed NATO records. The examination of a considerable number of NATO files returns a fresh narrative for the alliance's commitment to science and environmental protection. Yet many of the documents examined do not come from the NATO archive. Notwithstanding the admirable work of its archival team, which is currently promoting greater openness in scrutinizing the alliance's history, the limited releases prevent historians from interpreting key episodes.[37]

NATO's records possess another feature that is possibly even more problematic. Although any historical record group has omissions deriving from the selective disclosure of individual papers in the collection, those produced by the alliance's officers were written to *deliberately* omit reporting any item in meetings that overtly displayed disagreement. They were compiled in an effort to further strengthen the alliance by recording how consensus was reached and overlooking contrasts preventing it.

So the nature of the records currently hinders efforts to reconstruct the coalition's past in that they dangerously warp its interpretation. I have thus used fairly extensively records from other repositories in an effort to cast a different light on important episodes and reconstruct the conflict between leading administrators that official NATO records kept hidden.[38] I looked at different sets of archival papers not because of a special passion for arguments, but rather because I realized that the conflict was in fact decisive in sparking important changes in policy and diplomacy.

The problems encountered in retrieving the sources for this book are telling, more generally, of portraying NATO's controversial history while making sure that all the voices—of men and women, of supporters and dissenters, of major and minor powers, of leaders and low-ranking officials—resonated.[39] It is up to the reader to judge whether the book's narrative is pluralistic enough. Some will sense that US voices are too present, others that some within other NATO countries are not. I share these concerns, but the varied set of sources utilized shows, at least, good intentions.

The issues of pluralistic representation and sources call to mind a much broader concern about the construction of a satisfactory methodology to attack the problem of reconstructing NATO's science and environmental programs. In this respect, the growing momentum for a transnational history approach has informed my search for a suitable scheme, especially because of the book's focus on international scientific cooperation.[40]

This returns me to the question of archival sources and transnational methodology. While researching for the book, I realized that key decisions on NATO's programs did not take place just in one room, nor did they affect just one space. The interconnectedness of the stances of individual actors and national delegations meant that I had to visit the Italian national archive to find out about Norwegian stances on NATO's programs, and only after talking to a former science administrator in Norway was I able to fully understand what the Italians were really after when contributing to these programs. Transnational history presents remarkable challenges, but in completing this book, its merits appeared compelling to me.

It is important, however, to stress the limitations of this study—especially in light of the materials examined. The archival materials consulted are voluminous and detailed, the ramifications of NATO's programs are consequential, and the time span that this volume covers is broad, making this a preliminary survey of a much wider historical phenomenon. Moreover, while the coverage is essentially chronological, some periods are examined in greater depth than others. The treatment is uneven because the amount of declassified archival documents produced before 1970 is greater than that from following decades.

I add two final remarks to ease the reader's journey through the material. First, because of its role and emphasis on defense, throughout its history NATO has adopted very stringent document classification criteria. And although several documents have been downgraded, I believed it useful to cite them with the security grade with which they were originally produced. Second, NATO archival documents are in French and British English. Names of agencies and quotes from documents are therefore in their original spelling (e.g., "Programme" rather than "Program"; "Defence" rather than "Defense"), but the remainder of the text is in US English.

This book could not have been completed without the support of the European Research Council, which awarded me a grant to complete the project *The Earth under Surveillance* (research grant no. 241009). I must also thank all my collaborators in the project who have been a terrific source of inspiration: Roberto Cantoni, Sam Robinson, Néstor Herran, Lino Camprubí, Soraya Boudia, Leucha Veneer, Sebastian Grevsmühl, and Sebastian Soubiran. A special thank you to Matthew Adamson, who further analyzed this book in its draft stage and came up with extremely useful observations as well as broad-ranging ideas on how to improve it. I'd also like to thank Peder Roberts, who has been incredibly helpful in finding the right way to frame some of the issues discussed in the volume, and Doubravka Olšáková and Giulia Rispoli, whom I have worked with in recent science diplomacy projects. Two of these are worthy of mention in this context: the EU H2020 project "Inventing a Shared Science Diplomacy for Europe," InsSciDE (grant no. 770523), and the newly established Historical Commission on Science, Technology and Diplomacy in the context of the ICSU/IUHPST Division of History of Science and Technology. Also thanks to the two anonymous referees who provided extremely useful comments about the manuscript.

I have met many archivists who have been extremely sympathetic and helpful. The assistance of the NATO archivists Jozefina (Ineke) Deserno and Nicholas Nguyen has been outstanding. I also wish to thank other archive officers, including Eudes Nouvelot and Johannes Geurts (NATO Archives), James Peters (University of Manchester Library's archival collections), Adrian N. Burkett (UK National Oceanography Library, Southampton), Shelley (Charlotte) Erwin (Caltech Archive, Pasadena, California), Bridget Gillies (Archives and Collections Division of the University of East Anglia, Norwich, UK), Ryan Reft (US Library of Congress, Washington, DC), and Geppi Calcara (Archivio Centrale dello Stato, Rome, Italy). Silvia Filosa and Graziano Ferrari allowed me to study the archival papers of the Italian National Institute for Geophysics. I am also indebted to Nils Holme and Olav Blichner for offering their recollections about NATO. This book was completed at the Centre for the History of Science, Technology and Medicine (University of Manchester, Manchester, UK) and I am indebted to my colleagues there for useful exchanges. A visiting fellowship allowed me to complete some of its sections at the Max Planck History of Science Centre in Berlin, for which I thank Lorraine Daston and Elena Aronova. I am grateful to the organizers of several academic meetings, including John Agar, David Burigana, Pascal Griset, David Aubin, Janina Onuki, and Amanda Domingues, for casting light on some of the issues discussed in this book.

ONE
—

Setting the Atmosphere

Mr. Shinwell (Seaham, Lab.): Is it surprising that my right hon. friend has again changed his mind? Has he not been doing that for years?

Mr. Churchill: I have found that to move constantly and to change with environment is one of the best ways in which a man can put himself in relation with nature. (Laughter.)[1]

At the end of World War II, was the natural environment just a subject for repartee? The war-stricken European nations and a communist-dreading US administration had little time for ecological consciousness. Urban landscapes had been transformed by tremendous bombardments—in Dresden, Hamburg, London, and Coventry. The Japanese cities of Hiroshima and Nagasaki underwent devastating atomic attacks. Neither victorious allies nor defeated enemies had shown any respect for forests, fisheries, or livestock, nor were the flora and fauna a priority at the onset of the Cold War; a confrontation that the banter-prone British Prime Minister Winston Churchill identified with the Iron Curtain descending over the European continent.[2]

It is further evidence of this lack of interest in natural environments that by then the British ecologist Max Nicholson was busy broadening the "still narrow and precarious base" of nature conservancy organizations.[3] What in the late 1960s would blossom into the large-scale environmental movement had previously produced only elitist committees and corporatist political customs.[4] Even the word *environment* ambiguously defined a natural space as much as anything else topographical (which explains Churchill's joke).[5]

Yet in recent years a number of scientists had built their careers on studying the physical characteristics of natural environments. During World War II, meteorologists and oceanographers routinely found employment in military research organizations to explore how weather factors and sea changes affected operations. D-Day showed the importance of forecasting optimal weather conditions to land, which had been thoroughly examined before the Allied attack.[6] Of course, learning about the natural environment had been a key military task since ancient times, but the availability of new means of studying it, including radar and sonar, had dramatically changed research practices. As the electromagnetic and sound pulses produced by these instruments bounced off the layers of the atmosphere and the oceans, they probed clouds and currents, thus increasing knowledge on their characteristics.

After the war, the US administration, especially through its military agencies, became a prominent sponsor for environmental studies in recognition of their significance in the Cold War. It also solicited a similar investment from Western Europe, when the North Atlantic Treaty Organization (NATO) was established, as part of a broader investment to improve detection and communication operations. But these propositions found US allies in Western Europe passive, because they viewed NATO as a project to retain a monopoly on nuclear weapons while sharing coordinated defense. Support from the UK delegation at NATO eventually helped US officials to ram home plans for an investment in science through the establishment of the NATO Science Committee. Its meetings represented the first true exercise of science diplomacy within the alliance. Building a dialogue on scientific collaboration helped evade political divisions while fostering an investment in scientific research of significance to NATO's defensive mission.

United We Stand, Divided We Rearm

NATO was established through the North Atlantic Treaty signed on April 9, 1949, by the representatives of twelve nations (Belgium, Canada, Denmark, France, Iceland, Italy, Luxembourg, Netherlands, Norway, Portugal, United Kingdom, and the United States). Its main goal was to prevent and repel Soviet aggression. Article three of the treaty stated that defense was the alliance's *raison d'être*: NATO ought to "maintain and develop [. . .] individual and collective capacity to resist armed attack."[7] Article nine paved the way to the establishment and regular meeting of the NAC, which included high representatives of each member country.

Plans to establish a defense alliance dated back to 1947. The Berlin blockade, the dominating influence of the Soviet Union in Czechoslovakia, and Soviet threats against neighboring Western European countries such as Norway played a part in convincing US President Harry Truman to speed up its foundation, casting aside his soon-to-be allies' reluctance.[8] Even Truman, in fact, was wary of promoting full-fledged military and political integration. Mindful of the Monroe Doctrine, the traditional isolationist stance of US foreign policy, he originally perceived NATO as a loose mechanism for common defense, it being one of the few defensive alliances that the United States had concluded in its history.[9]

Truman pushed for greater integration when he realized that some countries in Western Europe (primarily Britain and France) would single-handedly rearm and seek to acquire nuclear weapons. Moreover, the re-armament of Western Europe seemed likely to require that of West Germany, bringing up the thorny matter of its postwar status, something that troubled the French especially.[10] And finally de Gaulle and Churchill were eager avoid playing the role of mere second-rank partners.

From 1947 on, Truman more openly renounced to the US isolationist stance, and the new doctrine of containment sought to counter Soviet influence in Western European countries through the economic assistance plan stipulated by Secretary of State George Marshall. In the final years of Truman's administration, assistance to economic recovery morphed into open support for the setting up of defense infrastructures and procurement of defense equipment.[11]

But differences within NATO on military and political coordination continued to hamper the alliance's activities. When in 1952 the NAC agreed to appoint a Supreme Allied Commander for the Atlantic (SACLANT), Churchill expected that the new NATO commander would be a UK officer and urged the US president to consider the key role played by British naval forces in World War II; the Atlantic sea "floor is white with the bones of Englishmen."[12] The appointment of Admiral Lynde D. McCormick as the new SACLANT represented a blow to Churchill's aspirations. It was not the last, as Truman agreed to accept Turkey's request to join NATO against Churchill's wish to integrate that state in a different military alliance. The US administration also recognized Greece as part of a strategic "outer ring" vital to Western defense. On September 21, 1951, the NAC approved the entrance of both countries into the alliance.[13]

Churchill had little time to register his complaints with Truman, for at the end of 1952 David Dwight Eisenhower was elected US president. Eisenhower promoted active containment and the new secretary of state, John Foster Dulles, articulated a novel approach, introducing the NATO strategic doctrine of massive retaliation. Encapsulated in NATO Military Committee document 48 (MC 48—approved by the NAC on December 17, 1954), it posited that collective security rested on the alliance's ability to launch devastating retaliatory attacks with nuclear weapons, the main deterrent to Soviet aggression.[14]

Because at this stage the United States was the only NATO country with nuclear weapons, the new doctrine anticipated the proposal to install nuclear missiles in Europe and the entry into service of the newest Strategic Air Command (SAC) nuclear bomber, the Boeing B-52. Medium-range (Jupiter and Thor) and tactical ballistic missiles were eventually offered to Italy, Britain, and Turkey. And when Eisenhower exited his office in the White House, the nuclear stockpile he left was twenty-two times as large as the one he had inherited.[15]

Meanwhile the British and French continued to fight "a losing battle for equality with their American colleagues in the Pentagon."[16] Plans to establish a European defense community increased tensions with France, and only West Germany's accession to the alliance, in 1954, restored some degree of unity.[17] If united the NATO allies stood, divided they rearmed. Italy and Turkey eventually agreed to host nuclear missiles, controlled through the "two-key" principle.[18] But Iceland's parliament opposed the presence of NATO naval bases. Norway and Denmark allowed them but wanted to have no nuclear weapons on their soil.[19] The stances of non-NATO European governments made the situation thornier. In 1953, Truman had already agreed to coordinate defense with the fascist dictator Francisco Franco in Spain, which frustrated the British even more.[20] Sweden boasted of its neutrality while gaining access to US technologies through the 1949 Mutual Defense Assistance Act.[21]

US approaches to national security unnerved European policymakers. In the 1950s the US Air Force used Swedish airspace in reconnaissance missions without Swedish consent.[22] Military buildup in Greenland was kept hidden from the Danes, so that when Denmark, which held sovereign rights, joined NATO, it "remained an ally with clear reservations."[23] In 1954, Secretary of State John Foster Dulles confirmed that European allies were "insisting that we are so militaristic."[24] Were they? Military spending by the US administration dwarfed that of European allies and in the last four years had steadily grown to three times as much as what the Europeans spent (see table 1.1).

Assistance to defense procurement soon became a key factor in appeasing local administrators and won over their foot-dragging in joining the US crusade against communism. Resistance to what government agencies in Washington, DC, proposed remained fierce regardless and would un-

Table 1.1 Military spending in North America and Western Europe.

	Military spending in North America	Military spending in Western Europe	Ratio
1950	145	81.7	1.77
1951	309	99	3.12
1952	437	124	3.52
1953	451	135	3.34
1954	390	129	3

Data in US$ (billions) at constant 2015 prices and exchange rates. Please note that aggregated data for North America comprises the United States and Canada and that Canada's military expenditure in these five years oscillates between 2 percent and 4 percent of US military expenditure. Source: "SIPRI Military Expenditure," https://www.sipri.org/sites/default/files/SIPRI-Milex-data-1949-2016.xlsx.

fold through proposals undermining the Atlantic alliance, including—for instance—plans for a European defense organization.[25] Furthermore, communist parties in France and Italy gained popularity (and votes) through peace campaigning and opposition to atomic weaponry. Leading European scientists such as Pierre Joliot-Curie led the campaign against the atom bomb (which targeted the newly born NATO as well).[26] The offer of financial assistance aimed to quell dissent and materialized the beginning of NATO's investment in defense science.

Flying Vanguards: AGARD and SHAPE

Exploiting the anxieties that the conflict in Korea produced, the US Air Force (USAF) took responsibility for sponsoring specific science and technology projects in Western Europe through its European Office of Air Research and Development Command (EOARDC). And two NATO organizations, the Advisory Group for Aeronautical Research and Development (AGARD) and the Supreme Headquarters of Allied Powers in Europe (SHAPE), propelled multilateral research for two complementary defense tasks of the alliance: the design of a new aircraft for tactical operations and a new warning, tracking, and surveillance system.

It was the director of USAF Scientific Advisory Board who first suggested a meeting of aeronautical research directors of leading NATO countries. It took place on February 5–9, 1952, in Washington, DC. After the meeting, British and French representative were persuaded to support the United States' call for pooling "scientific and technical skills, manpower and facilities" in order to establish "mutual cooperation."[27] These recommendations led to the establishment of AGARD, which eventually welcomed representatives from other Western European countries. But, as the name suggests, the advisory group's focus would be on aeronautics research and USAF funds paid for the group.[28]

Instrumental in the creation of this new NATO forum for the promotion of research was its first chairman, the Hungarian-born US mathematician and aeronautical engineer Theodore von Kármán. He later claimed that AGARD was conceived as "a pilot to test [the] feasibility" of scientific collaboration, whereas it was, in fact, an attempt to remove the prevailing reluctance of other NATO allies to invest more in coordinated defense.[29] Von Kármán's reputation loomed large in the Western scientific community and his swagger and occasional tease had helped him to gain a position of prestige in the restricted circle of scientific advisers in the US administration.[30] "Todor" had been regularly attempting to advocate new defense systems and projects since the end of the conflict and, in particular, he offered colleagues in southern Europe assistance in aeronautical research, for which his government was prepared to pay in some measure.

Former director of the Aeronautical Institute of Aachen (Germany), in the 1930s von Kármán became the head of the Guggenheim Aeronautical Laboratory at the California Institute of Technology (Caltech) of Pasadena and went on to establish the renowned Jet Propulsion Laboratory during World War II. Appointed a member of the USAF Scientific Advisory Board after the conflict, he provided direction for its fast-growing military complex. In 1950, he convinced the US secretary of defense that mobilizing the European scientific establishment was fundamental to Cold War competition. "The mobilization of science and creative engineering [. . .] will be an essential factor in determining success or failure," he wrote.[31]

AGARD was meant to work as a forum to let negotiations on defense research move into the domain of multilateral cooperation and fostered the signing of remunerative contracts for aeronautical research. The US Mutual Weapons Development Program (MWDP), which as far back as the Truman administration had accelerated the coordination of scientific and research tasks with Western Europe, funded most of the group's projects.[32] From the mid-1950s, US general Thomas B. Larkin directed the assistance program and stimulated research and development in Western Europe with hosting countries contributing 50 percent of related costs.[33] For instance, the MWDP catered to the acquisition of modern radar equipment for scientific research at the University of Ankara. In 1959, Turkey's AGARD representative stressed that the recently purchased state-of-the-art remote-sensing equipment had increased Turkey's preparedness for national defense.[34]

Eventually US sponsors succeeded in convincing allies in Western

Europe to invest more. They actually assigned funds and projects to one national group rather than the others to stir competition. So the Italian car manufacturer Fiat was given the lead in designing NATO's light strike fighter, and Turkish specialists pioneered radar monitoring at the European fringe. The Dutch led in the establishment of the Air Defense Technical Centre (SADTC) set up by SHAPE, NATO's operational center for defense operations, established in 1951. Set up at The Hague to work on "problems concerned with air operations, by aircraft or by missile, generally in Allied Command Europe" (ACE),[35] SADTC was funded directly by the USAF and contracted out to the Netherlands' National Defense Research Organization (RVO-TNO). It was divided into groups devoted to studies for early warning and surveillance. Under the RVO's chief executive, the Dutch physicist Gerardus J. Sizoo, SADTC evolved into a modern defense laboratory.[36]

SHAPE's growing commitment to defense research now provoked the appointment of a science adviser, and the US mathematician Howard Percy (Bob) Robertson took the post in 1954. A graduate of prestigious Caltech, Robertson was known in the US scientific community as a mathematical expert who straddled disciplinary divides. Between 1929 and 1947, he was doctoral assistant, associate, and then full professor of mathematical physics at Princeton University, where he elaborated a cosmological model consistent with his colleague Albert Einstein's theory of relativity.[37] But during the war, Robertson's interest in cosmology lost place to other pursuits, which he carried out in secrecy. Employed as scientific liaison between Britain and the US Office of Scientific Research and Development (the country's war research organization), he collaborated with British experts who were busy improving the accuracy of aerial bombing through operations research (i.e., the application of mathematical and statistical techniques to military operations).[38]

At the end of the war, Robertson rose through the ranks. In 1949, he was appointed to the Department of Defense (DoD) Weapon Systems Evaluation Group (WSEG) and ended up chairing the DoD Defense Science Board, thus taking responsibility for shaping the Truman administration's scientific research policy.[39] As foreign secretary of the National Academy of Sciences (NAS), chairman of the National Security Agency's Scientific Advisory Board, and adviser in the US Central Intelligence Agency (CIA), the man was quite evidently far too busy. Often unable to cope with the volume of mail in his inbox, he kept apologizing for not replying: "you should see the Everest of reports and proposals I haven't read."[40]

Robertson brought into NATO his own philosophy of modern warfare based on operations research, something that would position remote sensing as decisive to NATO's research investment.[41] He found military operations as fascinating as cosmological problems and had the same appreciation of the role that surveillance infrastructures played in war planning. He frequently compared a victorious army to a healthy human body: "its eyes are photo-reconnaissance, its ears the radio listening posts, its exploring fingers the secret agents sent out into the field [. . .] . Information from all these sources fed back into the brain, where the collating function of intelligence strives to understand the complete structure."[42] By the mid-1950s, Robertson would push for completing SHAPE's remote-sensing structures because of their significance for the alliance's surveillance tasks.

"The Atmosphere Where Flight Takes Place"

In 1953, an article on NATO's air defense contended that its "purely military components" were just the tip of an iceberg of radar warning and tracking stations, communication and data processing units, and control centers.[43] The "body" that Robertson had metaphorically described in his lectures was a whole continent with distinctive morphological features that SHAPE scientists had the duty to equip with sensory devices. As the 1953 article contended, "The NATO-Europe territory comprises an arc of almost regular curvature extending for 4,000 miles from the North Cape to the Caucasus. The NATO-Europe land area behind that arc is quite narrow; its most distant point is less than 1,000 miles from the Iron Curtain, and most of its vital areas are much less than 1,000 miles. [. . .] Ballistic missiles could traverse the distance in minutes."[44] Although massive retaliation could inflict major losses to Soviet forces, any Soviet attack ought to be promptly detected and communication channels defended. Modern radar stations soon filled the European space.

The number of research centers devoted to knowing more about the European environment grew too. Originally a derivative concern, research on the natural environment now took center stage in NATO's sponsorship plans paralleling a similar investment in the United States.[45] In 1953, the USAF Geophysics Research Directorate (Cambridge, Massachusetts) was established and, by then, the US Office of Naval Research lavishly funded oceanography departments in universities.[46] And the organization of the IGY, a coordinated set of research exercises around the globe due to take place from 1957 and encompassing the study of the oceans and the atmosphere, brought grist to the mill of those science planners who, like the geophysicist Lloyd Viel Berkner, believed that US scientists should take advantage of international cooperative efforts to learn more about the global environment.[47] Berkner's vision found support at the DoD too, especially when, in 1953, the electronic engineer Donald Quarles was appointed assistant secretary of defense for research and development. Quarles viewed the IGY as a coordinated worldwide effort that would yield basic information "to our national defense problems" especially that "relevant to radio and weather predictions and the properties of the upper atmosphere."[48]

Back at NATO, Robertson and von Kármán translated this interest in putting together environmental knowledge from their national domain to that of transatlantic scientific collaboration. The first AGARD general meeting had ended with the request that geophysics and meteorology feature among the sponsored disciplinary sectors.[49] The geophysicist Joseph Kaplan, who was to be the chair of the US IGY national committee, now urged von Kármán to organize a meeting of NATO geophysical experts to start a conversation on the physical properties of atmosphere and ionosphere. His suggestion now found European scientists sympathetic. For instance, the Norwegian scientist Leiv Harang, a pioneer of prewar and wartime studies on the effects of auroras on military communications at Tromsø's geophysical observatory, was very supportive of this proposition.

The US Weather Bureau's chief of scientific services, Harry Wexler, now took Kaplan's proposition forward and in 1956 the AGARD meeting *Polar Atmosphere* took place.[50] The meeting explored Arctic geography and climatology in relation to air and naval operations. This research focus was dictated by strategic needs, for nuclear-weapons-carrying SAC bombers routinely flew over the North Pole. From 1955, some of its bombers refueled in midair and the base of Thule (Greenland) was chosen as a location for the placement of nuclear ballistic missiles.[51] From the mid-1950s, the "High North" was also the location of key Soviet nuclear missile bases (the Plesetsk complex), nuclear-testing facilities (Novaya Zemlya), air force installations (Kola peninsula), and naval stations (Murmansk).[52]

In introducing the Oslo meeting, von Kármán's assistant Frank Wattendorf stressed the importance of the polar atmosphere for "military aviation and [for] advancing knowledge in the geophysical sciences, including weather prediction and control."[53] A young British meteorologist who had worked for the United Kingdom's air ministry before moving to Canada, Frederick Kenneth Hare, opened proceedings by discussing the features of the Arctic climate. Wexler recalled patterns and anomalies in the circulation of Arctic air as well as the characteristics of its stratosphere. Presentations on forecasting maritime cyclones (in connection with the movements of air and naval forces), the distribution of solar radiation (responsible for dangerous "whiteout" making snow-covered mountains invisible), and ice reconnaissance were also given. Charles C. Bates of the US Navy Hydrographic Office reported on techniques for ice forecasting.[54]

AGARD now invited a second group of experts to share ideas on the use of radar equipment to chart the properties of the ionosphere. The portion of the upper atmosphere that extends from sixty to one thousand kilometers above the earth's surface is responsible, through the scattering of radio signals, for their propagation at long distances. The meeting focused on the ionosphere's layering and irregularities, and on turbulence and disturbances in signal transmission.[55]

After these meetings, NATO's investment in key research areas with a distinctive military dimension grew. Its Standing Group (a subgroup of the NATO Military Committee comprising only US, UK, and French delegates) agreed to set up a meteorology committee in order to find ways to provide "more accurate forecast to greater heights over greater distances and for longer periods of time," especially through numerical weather prediction methods.[56] Meanwhile, AGARD set up an ionospheric research committee to look more closely at the "physical conditions in the atmosphere where flight takes place."[57] In planning this initiative, Robertson and von Kármán had been particularly successful in persuading their Norwegian colleagues, because in Norway a tradition existed in the physical study of the natural environment.[58] The Norwegian scientist Finn Lied, one of the speakers at the Oslo meeting, agreed to chair the new committee. An electrical engineer who graduated from the Norwegian Technological Institute of Trondheim, during World War II he had acted as adviser to the Norwegian military attaché, first in Stockholm and then in London. After the war, he took responsibility for the reorganization of defense research in Norway, setting up its national establishment (Norwegian Defense Research Establishment; NDRE) at Lillestrom.[59]

Sharing information on the physical properties of the atmosphere became even more important for the allies when SHAPE agreed to build a NATO-wide detection system. In the 1950s, the alliance already operated LORAN, a coordinated system for radio navigation and an emergency network based on telephone cables.[60] But from 1959 a detection system offering coverage to the whole of the alliance's territory was finally ready. The ACE High Scatter network united eighty-two stations in nine NATO countries and stretched from Norway to Turkey.[61]

It was designed to allow signals to be transmitted and received in NATO stations by bouncing them off the layers of the lowest sections of the atmosphere (troposphere) and the ionosphere. The ionospheric scatter system was administered through selected ACE stations in Oslo, Paris, Naples, and Izmir, whereas the tropospheric scatter system required a greater number of stations in Western Europe stretching from Senja, in northern Norway (the NATO military post closest to Murmansk), to Ankara and Pazar, in Turkey (the stations near Soviet enclaves in the Black Sea).

By the time ACE High was completed, other Western military agencies had set up similar scatter systems, including the Distant Early Warning (DEW) Line in Canada (just above the Arctic Circle) and the North Atlantic Radio System, which comprised five stations that transmitted signals from Western Europe to the DEW network via Greenland, Iceland, the Faroe Islands, and Britain.[62] Meanwhile, Lied's committee organized symposia on topics such as forward scatter communication (1957), ionization phenomena in the "E" (1958) and "F" (1959) layers, the absorption of signals in the ionosphere (1960), and disturbances of solar origin to communications (1961).[63]

US funds (via the MWDP) provided the additional resources needed to set up research facilities in Western Europe, especially studies focusing on ground environment and equipment for battlefield surveillance (radar, photographic, infrared, sound, and visible light devices).[64] Funds were distributed in northern and southern Europe alike, prioritizing surveillance of strategically vital areas. The Dutch astronomer Egbert A. Kreiken took responsibility for setting up a national observatory in Ankara and the cost of forty-two thousand US dollars. US funds for the device were made available because "knowledge of the ionosphere and upper atmosphere in the geographical area of Turkey" was "of fundamental importance to aspects of NATO Air Defense such as communication and long range detection."[65] The atmospheric physicist Michael Anastassiades (Greece's representative at the AGARD Ionospheric Research Committee) set up a new laboratory at the University of Athens.[66] Italian atmospheric scientists used EOARDC funds to explore ionospheric disturbances in the Northern Hemisphere.[67] On the whole, US funding facilitated the creation of laboratories devoted to environmental analysis in connection with NATO's surveillance operations.

While Robertson and von Kármán succeeded in promoting a dialogue with allies as well as ad hoc bilateral initiatives, they struggled to convince Western Europe's political leaders to invest more in NATO's science. Back in 1952, von Kármán had proposed to transform AGARD into NAC's scientific advisory board, but the council rejected his proposal.[68] Robertson had also solicited, without success, distributing fellowships to promising scientists on the grounds that from 1953 NATO had inaugurated a scholarship program.[69] But the alliance already sponsored cultural initiatives through its Committee on Information and Cultural Relations. In 1957, frustrated by NAC's opposition, Robertson let a colleague know that the council had "the wind taken out of our sails."[70]

When Robertson was about to leave his position as SHAPE science adviser, he feared that unless NATO could propel international scientific collaborations, allies would drift off toward preferring to promote their own national research programs or alternative international collaborative endeavors. These fears were partly justified; despite investing what was due in the alliance, key US allies such as France and Britain had spent far more in their own atomic energy research projects to promote an independent weapons program. Eisenhower's 1953 "Atoms for Peace" initiative had been conceived *exactly* to prevent similar developments in less-endowed countries such as Italy, the Netherlands, and Norway.[71] The establishment of the European Organization for Nuclear Research (CERN) the following year also aimed to promote cross-country collaboration, thus removing the risk of nuclear proliferation through independent national research projects.[72] Anxious about these efforts and eager to promote more transatlantic collaborative projects, Robertson thought about new ways to encourage allies in Western Europe to invest more. In particular, he and other US officials, including Isidor I. Rabi, his former colleague at Princeton, took advantage of NATO's diplomatic shortcomings to present such investment in a novel way.

Rabi and the Beginning of NATO's Science Diplomacy

In the 1950s, several US scientists, such as the MIT President James Killian and the physicist and celebrated radar pioneer Isidor I. Rabi, had worked (in combination the Ford Foundation and the CIA operative Shepard Stone) toward the building of an "invisible college" uniting scientific leaders in the United States and Western Europe who could, through the planning of new international projects, propagandize—as noted by Krige—quintessentially American values: "anti-Communist but not populist, nationalistic but not jingoist, firmly convinced that the United States, whatever its flaws, had its key roles in defending the Free World."[73] The language of science was, in the US propaganda effort, universal and heralding the same values as those propounded in liberal societies fighting against communism. The promotion of scientific collaboration was thus in itself a way to build bridges between nations through the establishment of scientific relations. It was also aligned with the principles of the 1950 "Science and Foreign Relations," a US Department of State survey on the diplomatic uses of science, for which the geophysicist Berkner and two colleagues, Phil Strong and Joseph Koepfli, were drafted. The report recommended investing more in foreign scientific relations, recognizing that science "provides an effective medium by means of which men can meet and exchange in an atmosphere of intellectual freedom and understanding."[74]

As one of the scientists involved in the initiative and the instigator of a NATO science program, Rabi had learned the trade of scientific advice during the war. A PhD graduate of Columbia University, he pioneered radar research at the MIT Radiation Laboratory. When the conflict ended, he returned to Columbia University (where he chaired the physics department). He was also made responsible for a number of government advisory committees. In 1949, as member of the Atomic Energy Commission's General Advisory Committee, Rabi famously collaborated with Enrico Fermi in writing the minority report against setting up a hydrogen bomb program; he later chaired the committee.[75] Rabi was as esteemed as von Kármán in Western Europe, and not just because his achievements in nuclear physics and radar research had led to him being awarded the 1944 Nobel Prize in Physics. In the postwar years, he worked toward making Marshall Plan funds available for the recovery of European research centers and helped establish CERN.[76]

Rabi was also a recognized public intellectual, speaking authoritatively about science and proselytizing about science and freedom in the context of anticommunist propaganda.[77] Rabi had also worked together with Robertson in the physics department at Princeton University and they were both members of Eisenhower's science advisory committee (together with Berkner and Killian).[78]

Rabi, Robertson, and Killian agreed that NATO should promote transatlantic scientific collaboration and started working on a variety of proposals to materialize a sponsorship program. But allies in Western Europe warmed up to the idea of a NATO investment only when they became more anxious about the possibility that the Soviets and their allies would prevail in the scientific and technological race with the West and therefore feared the consequences this superiority might have had for the prosecution of the Cold War. It was especially a recent publication on the lack of advanced scientific training in Western countries that persuaded them to start a conversation at NATO. In 1955, the USAF, the CIA, and the Carnegie Foundation sponsored a comparative study of investments on scientific manpower in the Eastern and Western blocs. Completed by the Russian-born scholar Nicholas DeWitt at the Russian Research Center of Harvard University, the study analyzed the levels of education and scientific literacy in the Soviet Union and produced apprehension in the Western world in the gap it envisaged in the training and recruitment of scientists. Published by the US National Science Foundation with the title *Soviet Professional Manpower*, it caused a sensation by materializing fears that a modernly equipped communist intelligentsia could supplant the outmoded Western intellectuals unfamiliar with recent advances in science.[79] It also brought to the fore the vexed question of what science policy was best suited to make researchers thrive in a time when the debate on freedom and planning of science was raging.[80] By the time DeWitt's work was published, the wave of consensus on the need for Western bloc countries to intervene in the problem of scientific manpower was mounting, and the NATO Conference of Parliamentarians (a recently established consultative structure) now appointed the Special Committee on Scientific and Technical Personnel under the chairmanship of US Senator Henry M. Jackson to make recommendations.[81]

The alliance's own diplomacy shortcomings further convinced political leaders in Europe about the merits of a NATO science program. When, in October 1956, Israel invaded Egypt, the French and British governments agreed to assist the Israeli forces without prior NATO consultation. Their intervention was far from successful and the Anglo-French support did not defeat Egypt's Gamal Abdel Nasser. With the Suez Canal still closed to shipping, US officials could further impress upon allies the need to strengthen NATO as a political entity. The NAC appointed the "Committee of Three" (or Three Wise Men), namely the foreign ministers of Italy (Gaetano Martino), Norway (Halvard Lange), and Canada (Lester Pearson), to analyze how to do so.[82] All three had important ties with US diplomacy. The Canadian Pearson had played a key role during the crisis in the Middle East as UN envoy and shared Eisenhower's stance for an immediate cease-fire. And the Italian Martino had been one of the key architects of US-Italy collaboration in science, especially atomic energy.

The Three Wise Men initiative placed for the first time the problem of science and technology at the center of NATO's diplomacy. Their report, which drew on the answers to a questionnaire distributed among national delegations, posited the importance of NATO cooperation in the information and propaganda fields and set a number of recommendations for better integrating the alliance.[83] It also recalled a once central tenet of the article "Science and Foreign Relations," stressing that science and technology could be decisive in improving economic cooperation and the security of nations. It further argued for the merits of such an investment to overcome economic underdevelopment and—hinting at the recent DeWitt's findings—stressed the need of recruiting and training more scientists. A parallel initiative of the NATO Conference of Parliamentarians recast the debate on scientific manpower, similarly advocating an effort to "increase the supply of scientists, engineers and technicians."[84]

The three diplomats' propositions on science and technology were now singled out in the NAC proceedings, and just two months after the Three Wise Men report was submitted, the council directed the setting up of a working group to look further into a NATO investment in science.[85] Robertson now sought to capitalize on the NAC deliberation to propose that such an investment would extend NATO's previous efforts through AGARD and SHAPE. He thus wrote to the director of Princeton's Physics Department, John Wheeler, that the US representative in the NAC working group should strongly advocate that NATO studies be "vital to defence."[86] He went on to bring Wheeler up to date, as he was chairman of a high-powered group of US experts tasked by Senator Jackson to prepare a preliminary report for the Conference of Parliamentarians about NATO's circumstances.[87]

This emphasis on studies vital to defense was not welcomed, however, by the European delegates, who advanced the Three Wise Men recommendations. When the working group met for the first time, the representatives of Canada, Italy, and Norway objected to this emphasis and, in light of their disagreement with the US official, the NAC could only solicit the appointment of a task force to better articulate a proposal.[88] Robertson's Caltech colleague (and coauthor of "Science and Foreign Relations") Joe Koepfli was now drafted to head the newly appointed task force.[89]

Although presumably Koepfli would have found considerable resistance to his plans, it was especially the UK delegation's support that made it easier for him to build consensus.[90] Such support echoed the growing synergies between the US and UK governments. If in 1956 French and British leaders had ignored the US administration in deciding to intervene in the Middle East, the situation had now changed dramatically. First, Britain's plans for acquiring nuclear weapons were reaching a critical phase, for a fusion device was to be tested in 1957. Second (and partly because of that), the British administration was about to secretly sign a mutual defense agreement granting access to US knowledge on nuclear weaponry. Moreover, sixty Thor missiles were about to be placed under the "two-key" arrangement on British territory.[91] And finally, in 1958 a special partnership would secretly bring together US and British officials negotiating a nuclear test ban in Geneva with their Russian counterparts.[92]

The restored commonality of intents meant that UK support was decisive in steering NATO's diplomacy decisively in the direction advocated by Koepfli. The task force's final document recalled that both problems of scientific manpower and cooperation required bold and imaginative actions. In particular, Koepfli now advocated the launch of the NATO Science Committee to coordinate activities in the fields of science and technology and impressed upon allies a sense of urgency about the need of setting it up. A dangerous imbalance existed, as the final document of the task force summarized, making it possible for the USSR to "outstrip the NATO alliance" by scientific and technical superiority.[93]

The recommendations may have raised more opposition at NATO. But the launch of the first Soviet satellite, Sputnik (notably a Soviet contribution to the IGY) strengthened calls to bring together NATO scientists to plan a future research program geared to defense imperatives. So when Koepfli's report was examined at the NAC Ministerial Meeting (the first at heads of governments level), on December 16–19, 1957, it received unanimous approval. US President Eisenhower led a delegation that included Killian and Koepfli.[94] A Science Committee would be established to "speak authoritatively on scientific policy" and a NATO science adviser would be appointed.[95]

The major political benefit to be derived from this collaboration was that allies would now be open to a dialogue, which in principle might have allowed them to overcome the difficulties emerging in tackling other issues, political and military, but in practice it advanced a US scheme for coordinating defense research so as to prevent, with UK assistance, its allies from independently developing more powerful (nuclear) deterrents. The further planning of collaborative initiatives would be left in the hands of the NATO Science Committee.

Special Partnerships

The Science Committee included national representatives who were all prominent science advisers in their national research organizations. From March 1958 on, they met twice a year at NATO headquarters to discuss the details of a sponsorship program. They were required to reach unanimity in the approval of new plans, which entailed much work in convincing reluctant national representatives. The science adviser and key members in the committee had often to persuade reluctant allies to fund the various programs, as recalled by the third NATO science adviser William Nierenberg.[96] But by 1958 US and British representatives were secretly working together to persuade other delegates about the need to provide an orientation to the committee's research grants scheme.

Actually, their friendly relations generated the much-needed consensus. It is important to briefly recall here that international scientific relations between Britain and the United States had developed significantly during World War II and that the conflict had molded personal relationships between officials who had been called on to advance specific national projects of military significance as well as to collaborate with colleagues. Research on atomic energy, radar, and strategic bombing had been particularly decisive in forging relations between US and British scientists, tightening their relations in the pursuit of particular military projects. At the end of the war, personnel in key roles were replaced, and information sharing in fields such as atomic energy abruptly stopped.

The organization of NATO's science program represented an opportunity to rekindle, as shown by Robertson's relationship with UK delegate Solly Zuckerman. A South African–born British scientist, "Sir Solly" is undoubtedly one of the most interesting figures in the history of contemporary British science.[97] A year older than Robertson, Zuckerman had originally trained in zoology and was interested in the study of the apes' anatomical features. As for many other scientists, the outbreak of World War II led him to deal with a number of other research tasks. Zuckerman had to cast aside his research interests and familiarize himself with operations research.

The forced detour led the zoologist Zuckerman close to the US mathematician Robertson, and during the war they often met at the High Wycombe RAF research bombing unit. When the war ended, Zuckerman returned to academic life with a bag of military honors and taught anatomy at the University of Birmingham. But he continued to advise his government as a member of the UK Advisory Council on Science Policy (ACSP), the main organism with oversight on postwar government investments in scientific research.[98] Meanwhile, his relationship with Robertson had grown into a close friendship. The two spent time in exclusive clubs on both sides of the Atlantic. Zuckerman introduced Robertson to the highly selective Athenaeum (the Royal Society fellow's club), whereas Robertson provided Zuckerman with a membership in New York's Cosmos, where they spent time "goofing."[99]

To Robertson's delight, Zuckerman was selected to represent Britain on the Science Committee. In 1957, the mathematician had visited London to confer with the anatomist.[100] Three years later, Zuckerman was appointed science adviser of Britain's minister of defence and Robertson congratulated him as a "happy combination of a research scientist, a man who has intimately grappled with the technical problems of defense, and one who has thought deeply of the relation between these two important spheres of endeavor."[101]

Zuckerman and Robertson joined forces in designing a convincing strategy for gearing the Science Committee's research program toward defense science. They were further assisted in their efforts by more familiar faces; for example, the man chosen as Robertson's alternate in the US delegation was the physicist Rabi. For sure, Robertson and Rabi represented the viewpoints of the DoD and State Department, respectively, in the context of the team of US officials (backstopping group) preparing statements for the committee meetings (including representatives from the NSF, the State Department's International Cooperation Administration, the AEC, and the White House). But they shared the ambition to set forward a program of fundamental research resonant with defense priorities.

The US candidate for the role of NATO science adviser was Harvard physics professor Norman Ramsey, a choice that could only be understood as manifesting the effort to further strengthen the synergies between Robertson and Rabi. Ramsey had successfully contributed to the radar program that Rabi had set up during World War II.[102] Robertson knew Ramsey, too, and in the postwar years they had been members of the USAF Science Advisory Board. In 1954, Robertson had warmly supported, without success, Ramsey's appointment to a high-level position in the National Security Agency.[103] It was thus Ramsey, Rabi, Robertson, and Zuckerman who took responsibility for generating consensus on NATO's science program. After the first Science Committee meeting, the amicable relations between "S.Z.," "Rabi," and "Bob" were impressed on photographic film just outside NATO's meeting room (figure 1.1).

Their convergence in thinking and planning echoed the improvement in political relations between their administrations. In contrast, it took much persuading to convince other committee delegates. In order to further assist Ramsey, Robertson recommended that he be assisted by the British officer H. A. (Tony) Sargeaunt, who had been his deputy at SHAPE. Sargeaunt's presence helped Ramsey to figure out ways to incorporate a defense research element in the Science Committee planning.

In March 1958, preparations for the first meeting of the Science Committee were almost completed; as if to mark the achievement, Ramsey invited Zuckerman to have dinner with Rabi, Robertson, and Koepfli the evening before the meeting.[104] These science advisers had by then already worked toward making sure that NATO's research investment resonated with the alliance's operational tasks. For instance, Robertson had prescribed the organization of an introductory course on operations research at MIT under the leading US expert in the field, Philip McCord Morse.[105]

One item these delegates had to be particularly wary of was information sharing. When the committee met for the first time, the thirteen national

FIGURE 1.1. Solly Zuckerman, Isidor Isaac Rabi, and Howard Percy Robertson at NATO headquarters. From Solly Zuckerman Archive, folder SZ/NATO/2, University of East Anglia (no identifiable author of the photograph).

delegates agreed to analyze existing information security rules and their implications for the activities of the new committee. Now Rabi formulated the proposition that the science adviser be entrusted with calling open and closed sessions as he deemed suitable to give him sufficient leverage in secrecy matters.[106] The Standing Group representative recalled that the basis of the alliance was "the need for collective defense" and the military was "the leading customer for the end product of scientific R&D."[107] Rabi thus intervened again to diplomatically stress the need to avoid making decisions on "extremely delicate" matters before the science adviser had conferred with "the appropriate agencies and individuals."[108] In this way, he gave Ramsey the opportunity to initiate consultations. Zuckerman supported Rabi's proposal: the committee should address defense matters, but the science adviser should decide what could be discussed.[109]

That the matter of secrecy would produce tensions was to be expected, especially because long and taxing negotiations had anticipated the debate on NATO science. Those on atomic energy, for instance, had reached conclusions only a few years earlier with the signing of the EURATOM treaty of March 25, 1957. But the treaty's security controls did not wholly satisfy US negotiators, who saw them as paving the way to nuclear proliferation.[110] Transatlantic negotiations were hampered by concerns on what information could be pooled without giving away critical details. Information security represented a stumbling block in the setting up of a joint NATO program also because of the existing US monopoly on nuclear weapons, something that would complicate transatlantic relations for several decades.

At its third meeting, the Science Committee eventually approved the new regulations on security. Rabi announced the release of need-to-know US reports in order to stimulate "essential defense research." Zuckerman followed suit, claiming that Britain's administration also intended to make available scientific literature to promote information sharing.[111]

Ramsey, with the assistance of both Zuckerman and Rabi, successfully geared NATO's science program toward a defensive agenda by introducing new information-sharing procedures in the context of the Science Committee. We shall see in the next chapter that what the third science adviser William Nierenberg called a "mixed philosophy" (i.e., of investing in scientific collaboration conducive both to strengthening relations between allies *and* enhancing defense) was further implemented when the committee members agreed on an investment in fields such as meteorology and oceanography.[112] A focus on the physical characteristics of the environment advanced the collaborative work originally set out by NATO's defense research organizations while a parallel arena for NATO diplomacy was now created. More diplomacy work and secret scheming was thus needed, especially in order to find allies prepared to support NATO's new research grants scheme.

TWO

The Surveillance Ambitions of NATO's Science Program

No doubt the [Science] Committee was helped into being by the rumbling of rockets and the alarming "bleeps" from the first Sputnik.[1]

Scientists interested in learning more about the earth and its features have always used a variety of instruments to acquire new knowledge. But in the 1960s, equipment transmitting and receiving powerful signals was the means of choice for a new generation of experts who had first used radio pulses in military operations during the war. As radio pulses bounced off the atmosphere, they returned information on its characteristics, on weather phenomena, on auroralike events, and much more. The innovative aspect of this scientific practice was its remoteness. The setting up of information-gathering facilities and the availability of sophisticated electronic eyes and ears made it unnecessary for the observer to be in direct contact with the object studied. Push-button benchtop practices and headphone-equipped listening posts replaced traditional fieldwork based on direct human observation, and the new age heralded the merits of remote sensing.

As this new knowledge was accumulated, experimenters and their patrons alike recognized its hidden potential for Cold War surveillance operations. The very same pulses used to chart sky and deep sea assisted in revealing, indirectly, what enemies were busy doing, especially in the atomic field. The testing of nuclear explosions would cause earthquakes and massive perturbations in the atmosphere, which could be remotely sensed miles away from the test location. Knowledge on how signals propagate in an environment also assisted in detecting the presence of enemy aircraft or spacecraft and vessels. Traditional surveillance work required the use of human agents in risky operations in foreign territories. On May 1, 1960, US pilot Gary Powers was famously caught collecting pictures of Soviet military installations while flying over them with a U-2 spy plane. The episode caused a media, diplomatic, and intelligence uproar, but actually his espionage mission was only one of many surveillance operations carried out by the US higher echelons. And the most profitable ones continued to be invisible. And by the time Powers was captured, NATO was already funding innovative studies with a surveillance focus. The alliance thus coupled US intelligence-gathering efforts with projects of the alliance's scientists that remotely sensed anything between the top of the atmosphere and the deepest reaches of the oceans.[2]

This chapter focuses on these surveillance ambitions. It looks first at NATO's sponsorship schemes and how the Science Committee debated research priorities. It then examines how the alliance's science adviser succeeded in implanting a "defense science" strand of funding into NATO's science program in order to prioritize studies tying to the alliance's defense and surveillance imperatives. We thus learn that the largest single award ever assigned in the context of the research grants scheme during the 1960s was for a project funding better detection of satellites' radio signals. And ionospheric scientists who had previously been active in the AGARD Ionospheric Research Committee now received additional funds for collaborative projects sponsored by the Science Committee. They also established a group to chart the impact of meteorological phenomena on the alliance's tracking and communication operations.

The second part of the chapter examines the activities of a NATO subcommittee devoted to oceanographic research. Of the five groups set up in the 1960s, this group received the largest amount of funds within NATO's research grants scheme for the set of projects that they elaborated. While recognizing the importance of oceanographic studies for a variety of human endeavors from forecasting to shipping traffic, the subcommittee's research tasks aligned its mission to naval surveillance and especially antisubmarine warfare.

The warping of NATO's science program marked the transitory dominance of US interests in the alliance's investment and the defeat of competing plans, such as the one elaborated by the French for a Western science foundation. It also defined the extension of NATO's science diplomacy through the creation of more research groups. In particular, sponsorship aimed to further international collaborative efforts initiated during the IGY while aligning them to Cold War political alliances.

A Statistical Overture

Published in 1967, *NATO and Science* documented the sponsorship activities promoted by the alliance's Science Committee in its first six years and widely disseminated its achievements. Assembled in the office of NATO's science adviser, the Scientific Affairs Division, the volume detailed the range and diversity of the alliance's science program. NATO had sponsored original research in a number of fields: the study of distant galaxies; hormonal dysfunctions leading to female virilisms; spores, fungi, and soil nutrients; and the botanical features of eastern Greenland. The book stated that most research grants were "modest in size" and targeted new areas "for fruitful international cooperation in research."[3] NATO's patronage thus appeared philanthropic, benign, and economically viable, dealing with fundamental research questions and neutral with respect to distribution by country and scientific disciplines.

But a statistical analysis of the data provided in *NATO and Science* reveals that the alliance's initial investment was distributed evenly only for its first two strands, whereas the third, consisting of grants for collaborative research projects, was less balanced. The first strand was to shore up scientific manpower by offering scholarships to young researchers. Because the proposal was approved by the NAC after the task force's report (and the Three Wise Men recommendations), the funding agreed upon (substantially higher than that of the other two strands) was promptly accepted by committee members. It kept growing from 1 million US dollars in 1959 to 2.6 million in 1966. (See figure 2.1.)[4]

The fellowship program enabled young scholars with an undergraduate degree (age twenty-six to thirty-five) to travel to prestigious academic institutions and work there for one (or more) years.[5] In the spring of 1958 the Science Committee appointed a working group from among its members to set out the criteria for awarding fellowships.[6] The first set of scholarships was assigned to fund students enrolling in 1959. From then onward the Science Committee took responsibility exclusively for recommending a budget for the program and let the national governments' research offices evaluate how to award the fellowships. US physicist John L. McLucas (NATO science adviser from 1964 to 1966) recalled a few years later that the program guidelines had been deliberately left loose; the content of specific research was "in the background of planning." (See figure 2.2.)[7]

A second strand, the Advanced Study Institutes (ASIs), aimed at letting prominent scientists from NATO countries organize workshops to discuss specific topics. While the Scientific Affairs Division took responsibility for applications, the initiative of organizing ASIs was left to the schools' "di-

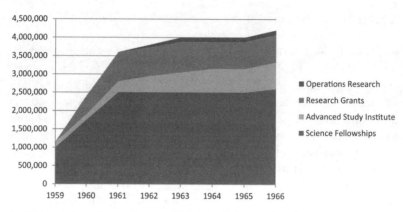

FIGURE 2.1. Annual expenditures on NATO Scientific Program by scheme. Data from NATO Scientific Affairs Division, *NATO and Science.*

rectors" (i.e., those scientists requesting funds to organize a workshop).[8] Committee members and officials at the NATO Scientific Affairs Division wished especially that directors take responsibility for publishing proceedings in order to give wider resonance to the ASI initiative.

The third strand of NATO actions to promote international cooperative research was different from the other two. The research grants program started later, in 1960, and was the subject of many debates at the Science Committee's meetings. When the committee finally approved it, five prestigious academics (Piero Caldirola, Italy; Hans Piloty, West Germany; Ewart H. Jones, Britain; Pierre Fleury, France; and H. W. Julius, Netherlands) were appointed to make recommendations on projects to be funded. By then, however, the committee had already appointed ad hoc working groups to stimulate international collaborative work in specific subjects of interest to NATO defense.

Three of these ad hoc working groups focused on oceanography, meteorology, and radio meteorology. Their members could make recommendations about specific projects to fund, thus receiving an overall budget for the groups' members and projects. They also received funding under a grant program aiming to boost international cooperation (A-type grants). Members from Greece, Turkey, Italy, Iceland, and Portugal also could apply for grants to develop national projects (B-type grants). Figure 2.3 shows the overall funding awarded to these groups, including projects proposed by the members of these three ad hoc groups in the A and B grant streams.

Between 1962 and 1964, projects with a focus on oceanography, meteorology, and radio meteorology received more than half of overall NATO

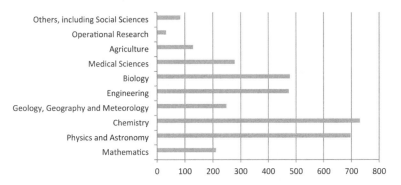

FIGURE 2.2. Subjects studied by fellows during the years 1959–63. Data from NATO Scientific Affairs Division, *NATO and Science.*

funds available in the grants stream (see figure 2.4). This emphasis was the result of a debate within the Science Committee in which propositions from US delegates, including Robertson, on promoting defense science received the support of other NATO allies. US and British "science diplomats" worked together in order to shepherd their colleagues toward the approval of a sponsorship agenda that accommodated defense ambitions.

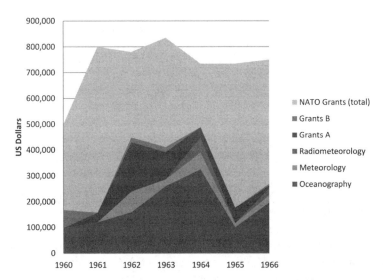

FIGURE 2.3. Data in US dollars on strands of the NATO research grants program. Data from NATO Scientific Affairs Division, *NATO and Science* elaborated by the author from *NATO and Science,* appendixes IV, V, VI, VII.

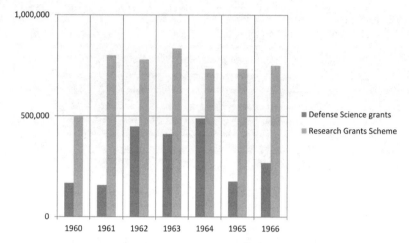

FIGURE 2.4. Cumulative funding of defense science grants (as shown in figure 2.3) as part of yearly budget for the NATO research grants scheme. Data from *NATO and Science*. Amounts in thousands of US dollars: in 1960, 168 out of 500 (34%); in 1961, 157 out of 800 (20%); in 1962, 448 out of 780 (58%); in 1963, 411 out of 835 (50%); in 1964, 488 out of 735 (67%); in 1965, 176 out of 735 (24%); in 1966, 268 out of 750 (36%).

Promoting (Defense) Science

Their plans urged them first to defeat a competing plan for a French Western science foundation, though. If the mutual support between UK delegate Zuckerman and the US representatives resonated with the positive moment in Anglo-American relations, the dissent between Zuckerman and the French representative was similarly rooted in the perils of the *entente cordiale*. After the Suez crisis, Anglo-French relations soured again and the establishment on January 1, 1958, of the European Economic Community further isolated Britain. With negotiations on a NATO science program ongoing, the French government had further advanced its atomic weapons program, and in September 1958 the CIA documented secret talks between foreign ministers of Italy, France, and West Germany about collaboration on the production of an atomic device.[9] De Gaulle attempted to reduce tensions with Britain and the United States by proposing, unsuccessfully, that NATO reorganize under a tripartite directorate. But now these political controversies affected the committee's proceedings too, especially when the French delegate André-Louis Danjon proposed establishing a NATO foundation for science.

Danjon, an astronomer, former director of the Strasbourg Observatory, and from 1945 head of the one in Paris, was another staunch supporter of defense research but hoped for a different administration of NATO research projects. Ramsey was wary, and Zuckerman openly against the plan.[10] In April 1958 an ad hoc group comprising Science Committee delegates from France, Norway, Italy, West Germany, and the United Kingdom met to discuss the French proposal. A French committee responsible for NATO scientific affairs was also established. Headed by the president of the socialist group at the National Assembly, it comprised the mineralogist Henri Longchambon and three prominent French physicists (Louis Néel, Yves Rocard, and Francis Perrin).[11]

Zuckerman now argued that a foundation would have left Science Committee members as mere overseers of a program developed elsewhere.[12] In the context of the British government organization responsible for national science policy, the UK Committee on Overseas Scientific Relations, he was even more vocal indicating the project as a "disguised French effort to get an institution contributing to the 'glory' of France, and which would give financial support to the pitifully penurious French pure science research effort."[13] At the next Science Committee meeting, Rabi supported Zuckerman in his attempt to kill the French proposal; it was wrong to assign to another body a duty that the NATO Task Force had reserved for the committee.[14]

The physicist Louis Néel eventually replaced Danjon as French delegate on the committee, and he let the plan for a foundation wither away. By 1960, Néel was one of the most influential figures in French science and had by then managed to set up a major complex for physics and nuclear research in Grenoble; in that year he proudly chaperoned General de Gaulle around the complex.[15]

Even with Danjon defeated, it took some time for Rabi and Zuckerman to persuade colleagues to endorse a sponsorship scheme that would be geared toward NATO's defense priorities, especially since they were equally unclear about what exactly needed to be prioritized. Tony Sargeaunt now suggested to Ramsey to call for a meeting of selected and highly qualified scientists representing the viewpoint of national defense organizations to get around other members' opposition.[16] So on December 1–2, 1958, NATO's "defense research directors" (DRDs) met for the first time. In introducing the meeting, the NATO Secretary General, Paul-Henri Spaak, stressed that its chief purpose was "to provide a forum in which those responsible for directing defense research in their own countries could express their views."[17] During the meeting a set of viable subjects for collaborative research emerged and aligned to NATO's previous investment in the physical environmental sciences.

The DRDs of Canada and Germany now reiterated the importance of studies on underwater acoustics, the propagation of signals in the iono-sphere, and geophysics.[18] The ex-chairman of the AGARD Ionospheric Research Committee (and now Norway's DRD), Finn Lied, recalled the need to "investigate all our geophysical environments in the widest sense and from a military point of view."[19] Frederick Brundrett of the UK Ministry of Defense stressed the importance of antisubmarine warfare.[20] US representative Robertson now recalled that geophysical studies could advance aerial navigation by offering "an improved understanding of the earth, its size, shape, gravitational and magnetic fields, inter-spacial bodies phenomena and solar manifestations."[21]

These opening talks had historical importance; eleven subjects were deemed suitable for scientific cooperation in defense science, and three received priority: (1) underwater detection, (2) oceanography, (3) upper atmosphere research.[22] Ramsey and his deputy Sargeaunt now called for a tailored research program.[23] During the Science Committee meeting that followed the DRDs', there was still resistance to Ramsey's plans. The Italian Francesco Giordani and other delegates continued to show reluctance to endorse the program, but, in contrast with Ramsey, they could not count on the support of defense authorities.[24]

Other problems threatened Ramsey's plan. The previous year, at an ACSP Committee on Overseas Scientific Relations meeting, a treasury official had manifested anxiety that the UK "should be paying the money and really not getting anything worthwhile in exchange."[25] As plans for a budget of a half-million US dollars for the NATO research program were about to be approved, Zuckerman promptly objected.[26] The members of the Overseas Scientific Relations Committee were persuaded only after a visit by Ramsey in March 1959, when the latter explained that the "there is reasonable prospect of the United Kingdom getting more out of the proposal than it contributes."[27]

By the beginning of 1960 there was still some uncertainty on what projects exactly the Science Committee should prioritize, especially with the US opposition to sharing information on nuclear weapons and missiles. On one occasion geophysicist Lloyd V. Berkner asked Koepfli and Rabi whether NATO could sponsor seismic research as, by then, Berkner had been appointed head of a government panel investigating what seismic equipment could be used to monitor nuclear weapon tests. Koepfli replied in the negative. Divulging information on seismic instrumentation would "necessitate disclosure of certain weapon characteristics."[28]

Not just for seismological equipment—competing views existed among US and European scientists in relation to orbiting satellites, despite the fact that it was Sputnik that had propelled the setting up of the Science

Committee. After the launch of Sputnik, von Kármán and the US physicist Fred Singer proposed to use a NATO research satellite for high-altitude meteorology (and to investigate the effects of shock and wave propagation of atmospheric nuclear explosions).[29] As chair of the US IGY National Committee Panel on Rocketry, Singer hoped to use satellites for environmental analyses.[30] AGARD's members were particularly supportive of a satellite scheme, and its deputy director, Eberhardt Rechtin (who had designed the US satellite Explorer), stressed that a NATO satellite would improve the study of the upper atmosphere's chemistry and ionization phenomena.[31]

But NATO scientists did not reach an agreement on a satellite project partly because of opposition to sharing information on vectors. Opposition from the UK delegation was decisive in preventing an agreement, and Zuckerman was instructed to avoid giving any indication about Rechtin's proposal, because it would disclose Britain's interest in acquiring a US missile launcher. At this point, the British government was about to cancel the expensive program for the Blue Streak vector and asked the US administration to supply Skybolt instead.[32] Von Kármán continued lobbying for a NATO satellite, though, and even tried to get the support of the Danish Nobel Prize–winning physicist Niels Bohr for his initiative.[33] Ramsey's successor, Frederick Seitz, eventually succeeded in arranging an "advisory group on space research." The meeting at NATO headquarters was attended by the Italian and Norwegian AGARD representatives Luigi Broglio and Finn Lied. But once again divisions emerged as the French remarked that "co-operation in space research under NATO auspices was not strictly indispensable."[34] The Italian was more supportive because an Italian satellite, *San Marco I*, was about to be launched with a NASA rocket.[35] But fearing disclosure of weapon technologies, US officials took the wind out of the NATO satellite's sails and suggested instead that a space research satellite be made available to the recently established International Council of Scientific Unions (ICSU) Committee on Space Research (COSPAR).[36] NATO collaboration appeared more feasible on the equally pressing issue of how to improve surveillance of orbiting satellites.

Spacetrack: NATO's Joint Satellite Studies Group

Signal propagation represented a vital component in the setting up of defense systems because the correct transmission of radio signals was decisive for purposes as diverse as the navigation of aircrafts and satellites, the identification and tracking of enemy forces, and the operation of advanced weapons systems such as missiles.[37] Under these pressing needs, US physicist Jules Aarons of the Geophysics Research Directorate (US Air Force Cambridge Research Center, AFCRC) succeeded in making of *Spacetrack* the focus of international collaboration under the NATO aegis. Aarons was already familiar with European research in the field of signal propagation. After being awarded a master's in physics at Boston University, a Fulbright grant allowed him to complete a PhD at the Paris-based Laboratory of Atmospheric Physics headed by the French rocket scientist Étienne Vassy.[38] Soon Aarons found employment at the AFCRC and further studied signal propagation, pioneering work on radio echoes from the moon for long-distance defense communication signals.[39]

Not only did Aarons further develop these studies at the AFCRC, but he succeeded in convincing a number of colleagues across the Atlantic, such as the British radio astronomer Bernard Lovell, to help. When Sputnik began to send radio pulses from space, Aarons suggested that the EOARDC fund research in Europe to track the satellite's signal at two European institutions. One was the Jodrell Bank Observatory (University of Manchester), where the largest radio telescope built to that date was about to be completed (see figure 2.5).[40] The other was the Microwaves Center (Centro Studi per la Fisica delle Microonde) at the University of Florence (Italy). Established in 1947, it drew on the tradition on research on signal propagation dating back to Guglielmo Marconi.[41] Before the war, the center's head, Nello Carrara, was recruited by the Italian navy. At the Livorno-based Italian naval academy, he had pioneered propagation studies. Shut down with the fall of the fascist regime, Carrara found new sponsors, including EOARDC, which helped propagation studies to flourish in Florence.[42] Personnel in Carrara's laboratory and at the Jodrell Bank had extensive experience in the use of radar-like apparatus and in the study of signal propagation and they were thus brought into the *Spacetrack* project to track signals from the satellite (code-named *Harvest Moon* at the AFCRC).[43]

In the second half of 1958, Aarons had already informed Lovell that the USAF was prepared to fund Doppler measurements of radio signals from the recently launched Sputnik II and III, thus furthering previous work on Sputnik I, in order to infer "the time of the closest approach of the satellite to the station."[44] In essence, Aarons was taking responsibility for finding more effective ways of tracking the satellites.

Eventually the radio scientist proposed monitoring other satellites (the recently launched Lunik) in the context of a broader research program and extending the collaboration to other European institutions such as

FIGURE 2.5. Lovell radio telescope under construction at the Jodrell Bank Observatory. Source: Jodrell Bank Centre, University of Manchester.

Carrara's and Vassy's laboratories in Florence and Paris, the Norwegian NDRE station of Kjeller, the Ionospheric Institute of Bresiach (West Germany), and the Kiruna Geophysical Observatory in Sweden.[45] Some of these institutions already had bilateral programs in place, as from 1952 Vassy and the German ionospheric expert Karl Rawer, head of the institute in Bresiach, had used the Véronique rocket to probe the radio transmitting properties of the atmosphere.[46]

The collaborative exercise demonstrated the benefits to be derived from tracking the same satellites at different latitudes and longitudes. In addition, it led to the discovery of a variety of unknown environmental phenomena obstructing signal reception. Once they enter the atmosphere, satellite signals' bend and warp. Their fading and, at times, disappearance suggested interactions with the atmospheric medium. Aarons now started to consider whether NATO could fund the expansion of *Spacetrack* work.

When Aarons realized that a NATO satellite project was never to take off, he considered asking funds for the consortium of laboratories busy tracking Sputnik III. By then the DRDs had already indicated (and the Science Committee approved) that the study of the upper atmosphere should receive priority in NATO's research grants program. In June 1961, Aarons visited Florence and suggested that Carrara prepare an application for a NATO grant on behalf of the *Spacetrack* consortia. With funding from EOARDC running out (and AGARD unable to take over because of its terms of reference), applying for the funds made available through the Science Committee appeared to Aarons the only way to extend the surveillance exercise.[47] These plans further developed when the Greek physicist Michael Anastassiades organized the NATO Advanced Study Institute on the Effects of Astronomical and Meteorological Factors on Wave Propagation for Radio Communications in Corfu (Greece, August 6–20, 1961). As we have seen, Anastassiades was the head of a recently established ionospheric institute at the observatory of the National Technical University of Athens. He now invited Carrara, Vassy, and Aarons to take part in the proceedings.[48]

The international character of Aarons's proposal, united with its surveillance agenda, made it a winner. The collaborative project received the largest Science Committee grant awarded in the first five years, including the largest individual grant in one application ($107,000 in 1962), and the largest grant through follow-up applications ($154,000 between 1962 and 1965).[49] Carrara's proposal of September 1961 openly posited its surveillance ambitions: "After the launching of the first Soviet Satellites, an effort was made by the USAF to MAINTAIN CONSTANT SURVEILLANCE of 'objects in space.' The Propagation Sciences Laboratory [. . .] undertook to ask several of the laboratories in Europe already under contract for propagation [. . .] in the effort both to keep track of the position of the transmitting satellite and to investigate propagation effects."[50] In the grant proposal, information on satellites that could be put together in this way included a study of their "reaction capability," their position, and, in some cases, "the point of release of the final landing stage."[51]

The exercise instigated novel research on how the earth's atmosphere facilitated the propagation of radio signals. In 1962, Anastassiades secured funding for an ASI symposium on radio astronomical and satellite studies (again in Corfu).[52] During these years, the representatives of the satellite surveillance group met regularly and set new tasks, such as studying variations in the electron content of sections of the atmosphere by comparing the fading of satellite signals or by measuring fluctuations in the intensity of signals, especially in polar regions.[53] This is indicative of both the influence that AGARD-affiliated scientists could yield in the sponsorship activities of the Science Committee and of the opportunity that this patronage offered them when AGARD could not fund their projects.

Reconnaissance Science behind the "Outer Ring": The Pursuit of Radio Meteorology

That AGARD scientists were more frequently invited to apply for NATO grants was no secret. For instance, when the Italian aeronautical engineer Broglio understood that his proposal for a study of the weather and climate patterns in Sardinia (where the new missile range in the area of Salto di Quirra was about to be set up) would not be funded by AGARD, he inquired through Francesco Giordani, the Italian representative at the Science Committee, whether it was possible to obtain a NATO research grant.[54] The head of the Italian Institute for Geophysics, Enrico Medi, also was confident that he could get funds previously made available by NATO through national defense agencies. In May 1960 Medi wrote to both Seitz and the Italian member of the Advisory Panel on Research Grants, Piero Caldirola, asking for funds to set up a network of stations in Italy for collecting data on the ionosphere. The proposal was first sent to General Thomas Larkin and then to the NATO Scientific Affairs Division.[55]

AGARD scientists played a prominent role in the organization of NATO's science program as recipients of both grant A (international collaborations) and grant B (national efforts). Vassy (who succeeded Lied as chair of the AGARD Ionospheric Research Committee in 1962–64) and Anastassiades (who succeeded Vassy the following year) used a grant of twenty-two thousand US dollars to further the collaboration between their laboratories in Paris and Athens.[56] Ten thousand more was given to Anastassiades for collaboration with Paris and Istanbul on a study of meteorological factors affecting radio-wave propagation. The head of the Ionospheric Institute of Bresiach, Karl Rawer, received a grant of forty-four thousand US dollars for an exploration of the E-layer in collaboration with the (lavishly funded) Athens' observatory. Lied was granted seventy-five thousand more to look instead at the D-region, always in collaboration with Anastassiades. And a further fifty-five thousand was awarded for studies on the index of refraction in the troposphere. The newly established astronomical center in Ankara also received funds for research in collaboration with *Spacetrack* (twenty-one thousand) and for the study of the E-layer (seven thousand).[57] Both Anastassiades's observatory in Athens and the new institute in Ankara headed by Egbert Kreiken could gain additional funds through B-type grants, which Anastassiades used for equipping the institute with a panoramic ionospheric recorder and a radio telescope (forty-one thousand in two grants).[58]

One reason that the heads of the Athens observatory and Ankara institute obtained NATO funds was that these facilities represented key surveillance outposts in what the alliance's military authorities identified as NATO's "outer ring"; its key geographic space for detection and tracking operations. The Greek observatory was a very important listening post where a variety of monitoring techniques could be used to gain information on Soviet activities, especially in the nuclear field. Anastassiades received funds for monitoring atomic tests by atmospheric means when the largest atomic bomb ever tested by the Soviets, *Tsar Bomba*, was detonated on October 30, 1961. The Greek scientists examined its ionospheric effects and one of Anastassiades's assistants received funding to collect data on the level of radioactivity in the atmosphere.[59] He was not alone. Leiv Harang, the Norwegian contributor to the 1956 *Polar Atmosphere* symposium, received sponsorship for a joint study between the universities of Oslo and Paris to conduct spectral analyses of luminosity in the night sky resulting from "explosive detonations in the ionosphere."[60]

The stream of funding that these experts secured grew wider thanks to another initiative agreed upon at another ASI workshop in Corfu. In summer 1961 the participants proposed that the Science Committee establish an advisory group devoted to radio meteorology. This umbrella term encompassed studies aimed at improving weather forecasting by radio techniques and radar equipment and at examining the impact of weather and atmospheric factors on communications. At the end of that year, the Science Committee approved the establishment of a radio meteorology advisory group.[61] The group comprised the ubiquitous Anastassiades with representatives from the French Centre National d'Études des Télécommunications (CNET) and the Technical University of Istanbul. It was chaired by the Italian meteorologist Luigi Santomauro (Brera Meteorological Observatory, Milan), who had previously received funding from the Italian air force to study meteorological factors affecting signal propagation in the Mediterranean region.

Having by now been appointed as the Greek representative, Anastassiades himself put the proposal to set up a new group before the Science Committee. Although the radio meteorology group aimed to collect environmental knowledge, this research task aligned to specific defense problems, especially from 1963, when the group expanded and the German meteorologist Karl Brocks of the Geophysics Institute of Hamburg agreed to chair it. He proposed that the group focus on the meteorological factors affecting transmission links across NATO countries. Since the SHAPE early warning system ACE High was now operational, the radio meteorology group agreed to chart the fading of its signals arising from weather conditions in areas surrounding key radio links connecting France and Turkey (those between Nice and Pisa, Athens and Smyrna, Martina Franca and Corfu).[62]

A (Well-Known) Cosmic Top Secret: NATO's Oceanography[63]

As figure 2.3 shows, the subcommittee on oceanographic research was the Science Committee's subgroup that in the period 1960–65 received the largest portion of funds disbursed through the research grants scheme. This investment in oceanographic studies was rooted, as that on atmospheric studies, in surveillance ambitions. By the late 1950s, the threat of Soviet submarines had become more visible and the development of the submarine-launched ballistic missile (SLBM) was a game changer in Cold War history. On January 7, 1960, a missile that could be installed on submarines, Polaris, was tested at the base of Cape Canaveral (Florida). The US navy had started research on a vector that could be launched from submersibles after collecting intelligence indicating the contemporaneous production in the Soviet Union of a diesel-electric-powered submarine equipped with nuclear missiles.[64] Polaris was promptly deployed on nuclear-powered submarines of the Nautilus class and from 1962 the British navy adopted it too.[65]

Newly elected US President John Fitzgerald Kennedy dramatically changed the investment in defense partly because of a new emphasis on naval warfare. As historian Walter A. McDougall has argued, Kennedy's administration heralded the beginning of an age of technological optimism in which funds were made available for far-reaching projects, including that of landing a man on the moon. Meanwhile, Robertson and his colleagues at the DoD Weapons Systems Evaluation Group reconsidered the mix of nuclear weapon delivery systems, now including the Polaris missiles, while the budget for foreign science programs was reduced.[66]

The new emphasis on naval warfare meant renewed attention to the study of oceans and their physical properties. This was something that Western intelligence agencies had monitored especially during the IGY, when Soviet oceanographic investigations had been one of the key features in their national program. So in 1959, a UK Ministry of Defence report revealed that Soviet oceanographers had carried out surveys in the Atlantic corridor and looked into its characteristics through radio controlled buoys.[67] Although a Soviet sea threat existed, Western naval intelligence exaggerated it. The 1950s Soviet fleet of ballistic-missile-carrying submarines NATO-coded as *Zulu* and *Whiskey* could only deliver the payload from the water surface. The nuclear-powered *November*, which began operation in 1958, was much noisier than those in the *Nautilus* class and therefore less suitable for strategic operations.

The changes in the strategic theater persuaded NATO's military authorities to upgrade the means to patrol and monitor sea waters and invest more in oceanographic research.[68] Underwater surveillance systems were by and large based on sonar devices, which detect the presence of enemy vessels by recording the sound they produce (or sound pulses that, as with radar, bounce off vessels when emitted by a source). And if radio propagation in the sky was affected by the properties of the atmosphere, that of sound was affected by those of the seas they traveled within. In particular, signals bent in sea currents from variations in water temperature, salinity, and density thus preventing surveillance, or—even more worrying—letting the passage of menacing enemy submarines hiding in currents go undetected. The biggest menace for Western naval forces was not represented by the enemy submarines as such but rather by the "layer effect" warping the sound waves away from potential targets and creating a shadow zone where the submarine could hide.[69]

Western navies invested in oceanography too, but they lacked coordination. During the IGY, two prominent oceanography centers, the Woods Hole Oceanographic Institution (WHOI) in the United States and the UK National Institute of Oceanography (NIO), had worked together mapping the Atlantic currents. The IGY provided an excellent opportunity for their oceanographers to complete this study of water circulation and start one on the Mediterranean.[70]

Already in 1956 a research group from MIT had placed underwater hydrophones off the US coasts in the context of what was called Project Hartwell. During the 1950s, the system expanded significantly under a follow-up project, eventually constituting what would be the core of the US Sound Surveillance System (SOSUS). This paralleled the provision of active sonar systems to Western navies to monitor the transit of vessels.[71]

In October 1958 the NATO SACLANT, US admiral Jerauld Wright, stated in a speech addressed to the NAC that a specialized NATO center devoted to antisubmarine warfare should be established. When the Italian government offered to make available disused buildings in the port of La Spezia (Liguria), the NAC approved plans for establishing the research facility, which was operational by 1959. A research ship also was made available: the converted two-thousand-ton former merchant freighter *Aragonese* (see figure 2.6). The DoD initially took responsibility for SACLANTCEN funding through a MWDP contract of 1.6 million US dollars annually. A commercial vector, SIRIMAR, administered the center.

Opened on May 2, 1959, the center was initially directed by the US physi-

FIGURE 2.6. The freighter *Aragonese* leaving docks in La Spezia (Italy). Source: NATO Archives, undated.

cist Eugene Booth, who during the war had been working in Rabi's physics institute at Columbia University. Specialists from nine NATO countries (Canada, Denmark, France, Germany, Italy, Netherlands, Norway, United Kingdom, United States) moved to SACLANTCEN to work on classified and unclassified research, although applicative antisubmarine warfare work received priority in its program. In line with its terms of reference, the center was divided into three groups looking into the production of new sonar equipment, the application of operations research to naval tasks, and oceanographic studies.[72]

Recognizing the importance of learning more about the sea environment to advance crucial sea surveillance tasks, SACLANT now lobbied for more support. And the strategic implications of a concerted NATO oceanography effort were divulged during the second Science Committee meeting, when the SACLANT representative, US Captain Kenneth M. Gentry (chief of the Naval Operations office) spoke to delegates about NATO plans for sea operations (the speech was later archived as a cosmic top secret; the highest level of NATO sensitivity).[73]

The captain explained that at the outbreak of a conflict, the bulk of the Soviet submarine fleet in the Arctic, Baltic, and Black seas would head toward the Atlantic. Naval forces would thus attempt to block the convoys before they reached the ocean in order to limit their operations.[74]

Consistent with the principle of massive retaliation, the SACLANT representative explained that NATO navies would ready for "destruction at source" of enemy vessels (eliminate the threat in their ports of exit: Leningrad, Murmansk, and Sevastopol—with nuclear weapons if necessary). Because the vessels, and especially submarines, were unlikely be found at their bases, however, Western navies needed to improve their "detection and kill capability" in sea areas providing access to open waters.[75] Gentry's focus on choke points and narrow waterways meant that it was especially the knowledge of sea areas around the detection lines cutting across seven passages leading to the Atlantic that needed to be put together in order to turn "detection into a kill."[76]

Gentry's words, united with the knowledge that soon a NATO center specializing in antisubmarine warfare would be set up and the DRDs' promised support for oceanographic research, persuaded Science Committee members to prioritize these studies. Ramsey thus contacted a number of Western oceanographers, including those who had contributed to IGY activities and played a key role in the establishment of its legacy organization; ICSU's Scientific Committee on Oceanographic Research (SCOR).[77]

SCOR's first official meeting had taken place in 1957 at Woods Hole Oceanographic Institution, when its director Columbus Iselin was elected chairman.[78] NATO's sponsorship efforts had thus both defense and political ambitions; hoping to take the research themes SCOR intended to pursue within the context of existing political and military alliances, and thus reducing the influence that cross–Iron Curtain collaboration could play in international oceanographic research. It also aimed to prevent the unwanted disclosure of oceanographic information—especially since the Soviet oceanographer Lev Zenkevich was a SCOR member—while pooling resources and knowledge among NATO allies.[79]

Ramsey first approached the Dane marine biologist Anton Bruun, one of the SCOR leaders who was extremely knowledgeable about Soviet oceanography.[80] Bruun and Ramsey eventually agreed to invite the leading SCOR oceanographers to NATO. Iselin had pioneered the study of the Gulf Stream and had been instrumental in the study's further development during the IGY.[81] His closest collaborator, the NIO director George Raven Deacon, had pioneered the study of sea currents too. They had both been active in World War II antisubmarine work when the latter headed British Admiralty's Group W (Waves).[82]

Iselin and Deacon's career trajectories show a much deeper synergy, in that they represented a tradition in marine science often referred to as "dynamic" or "physical" oceanography. Stemming from the work of Norwegian Vilhelm Bjerknes at the Geophysical Institute of Bergen, it informed oceanographic studies, forging a new generation of scholars.[83] The third expert invited to the meeting, the Norwegian Håkon Mosby, was a prominent member of the same school. A veteran of oceanographic expeditions, the Bergen-based Mosby had, like Deacon, specialized in the study of currents and sea temperature layering before World War II and was familiar with the use of these studies in surveillance and detection work during the conflict.[84]

On February 25–26, 1959, the experts met at NATO headquarters in Paris. Henri Lacombe of the Paris-based Laboratory of Oceanography, a "rising star" in French marine science, also was invited.[85] The experts proffered the words that Ramsey wanted to hear. Aware of the differences and variety of interests within the Science Committee, they did not emphasize exclusively the military implications of oceanographic research.[86] But their wish to dovetail the new SACLANTCEN activities was nonetheless made clear. Deacon openly admitted the defense implications of oceanography on the occasion of the first meeting of SACLANTCEN's advisory council, with Rabi also in attendance.[87]

A few days later, Ramsey's deputy Sargeaunt summarized these surveillance ambitions in a letter to Deacon. Western oceanographers would receive NATO funds to investigate the physics and layering of oceans in the same way in which ionosphere scientists looked into that of the atmosphere to improve telecommunications and detection.[88] A year later the Science Committee delegates traveled to the SACLANTCEN, where they held their meeting and Rabi recalled that on that occasion the United States accorded "very high priority to oceanographic research."[89]

Because of the mounting interest in oceanography, the NAC eventually assigned to the committee's subgroup the status of committee, which granted it greater freedom in the allocation of funds.[90] The NATO subcommittee on oceanographic research met for the first time in New York and starting from 1960 invested a sum close to two hundred thousand US dollars per year; nearly four times as large as SCOR's yearly budget.[91] This investment aimed at addressing the lack of knowledge on the seas surrounding Europe that Gentry's speech had indirectly revealed. The oceanographers agreed to carry out surveys on two sea areas of vital interest for the surveillance of Soviet vessels: the Strait of Gibraltar and the Faroe-Shetland Channel. Starting from 1960, the NATO funds were made available to complete these surveys also, thanks to the deployment of research vessels from the participating nations: the British *Discovery*, the Norwegian *Helland-Hansen*, and the French *Calypso*—the last famously utilized by Jacques Cousteau. SACLANTCEN's vessel *Aragonese* also contributed.[92]

The Norwegian Mosby was nominated as subcommittee chairman, a responsibility that he retained up until 1965. He also took responsibility for surveys in the North Sea, since he had previously looked into the dynamics of Norwegian sea currents. NATO funding thus helped him to explore further the Faroe-Shetland Channel.[93] It is important to note that these studies took place as the Norwegian stations of the US radio navigation system Loran-C were being set up; the system was a vital aspect of the Polaris program and important for NATO communications.[94]

Mosby's colleague Lacombe took responsibility for the study of Gibraltar and the adjacent sea area around the island Alboran. The study of this area was particularly important in that it allowed to put together critical evidence on the movement of currents in and out the Mediterranean. Gibraltar had been the focus of both oceanographic work and surveillance operations for centuries. But the study of currents had always proved challenging because of the considerable variation in water depths and the dynamics of water exchanges with the Atlantic. The spotty research French and Spanish oceanographers carried out during the IGY was now continued with the assistance of up-to-date instrumentation. This work shed new light on Gibraltar's distinctive movement of currents. As Camprubí and Robinson have recently argued, the study helped to visualize for the first time the idea of a global interconnected ocean, especially insofar as the modeling of seawater movement assisted in understanding other key chokepoints and current flows on the planet.[95]

The oceanic survey was as pioneering as it was divisive for the allies. Oceanographic exercises enhanced surveillance capabilities but proved taxing for NATO diplomacy. In particular, Lacombe invited the Spanish oceanographer Nicanor Menéndez García to contribute to the set of NATO trials arranged in the Mediterranean at a time when Gibraltar's sovereignty was disputed. This put the British Deacon in a difficult position at home in light of the UK government's avowed enmity to Franco's regime.

The NATO study now morphed into a complex diplomacy exercise, especially because US officials attempted to play down British resistance to Spanish participation.[96] Diplomacy was equally relevant to the other NATO oceanographic survey in the North Sea, where the strategic need of patrolling sea areas (also through oceanographic surveys), clashed with the economic interests of fishermen and fishery scientists. This was of marginal importance to Mosby and Deacon, who were busy developing a new type of automatic current meter so that a system of moored buoys for current recordings in the Faroe-Shetland Channel was laid out. The engineer Odd Dahl of the Christian Michelsen Institute in Bergen took responsibility for building the new current meter and recruited the Norwegian Ivar Aanderaa to design it. Funded at fifty thousand US dollars per year, it was eventually produced as a commercial prototype (the Aanderaa RCM4) by the British firm Plessey.[97]

But the surveys that the automatic current meter helped to complete overlapped those organized by the International Council for the Exploration of the Sea (ICES). One of many international oceanographic organizations devoted to sea studies, the ICES was distinctive for two important reasons. First, Soviet scientists were among its members—while US scientists were not. Second, it had the interests of fishery scientists, rather than national navies, represented within it. So NATO's effort to survey the area appeared to ICES members unnecessary; disrupting collaborative work with Soviet fishery experts.[98]

NATO's oceanography undoubtedly put surveillance before diplomacy. In the case of one survey of the Turkish straits, not only did the survey hit on one of the critical choke points that Gentry's speech had identified in 1959, but it complemented special gravity and magnetic measurements jointly carried out by the US Office of Naval Research and the Turkish navy's hydrography agency.[99] In the case of the Irminger sea survey (another vital transit for submarine warfare, in that it coincides with what navy strategists define as the Greenland-Iceland-UK Gap), Deacon dissuaded an Icelandic oceanographer from carrying out a similar study of another passage, that between Scotland and Iceland, since it had importance for fishery studies but not for surveillance.[100]

The NATO oceanographic subcommittee also contributed to the design of moored structures containing recording devices to collect oceanic data. These had value as scientific installations but also ensured the collection of intelligence on the transit of ships (and submarines) across passages. Buoy installations offered ideal cover for surveillance, especially since the SOSUS ensured prompt detection only as far as the eastern Atlantic. The British admiralty invested significantly in designing new types of buoys for the identification of submarines.[101] And the Soviet navy retaliated by installing its own buoy systems of surveillance (and their trawlers would sabotage or steal Western allies' buoys).[102]

That said, in the first half of the 1960s, NATO-funded oceanographers enjoyed considerable clout; something that the appointment of US physicist William Nierenberg as science adviser further demonstrates. Yet another of Rabi's students, Nierenberg had played a key role in the advancement of oceanographic studies and underwater detection during World War II. Nierenberg's year as NATO science adviser coincided with the golden age in the sponsorship of oceanographic, atmospheric, and ionospheric studies.[103] Just to limit our analysis to the subcommittee on oceanographic research, when, in 1965, the subcommittee completed its first summary report, it had designed and approved twenty-two projects, employed sixty-five scientists in the establishments of the participating institutions, and published nineteen technical reports. Between 1960 and 1964, it received funding of seven hundred thousand US dollars, to which national organizations added a further three hundred thousand.[104]

The stimulus to such an investment derived essentially from an understanding of connections between science and surveillance that was grounded on a philosophy derived from Robertson's teaching. This approach emphasized that an accurate understanding of the natural environment in which signals traveled, whether the sea or the atmosphere, was a decisive item in NATO's defense agenda. Ramsey, with the assistance of Robertson, Rabi, and Zuckerman, had paved the way to a defense science program. The ambitions behind such a sponsorship were kept hidden, although its merits for strengthening the alliance were divulged more widely.

Yet we have seen in this chapter that the launch of a mixed-philosophy program geared toward a defense agenda eventually made it difficult for its promoters to fully expound its merits in diplomacy. This was the case in radio meteorology and satellite studies, since preexisting transatlantic scientific relations (established via AGARD) could now flourish. But this was not entirely the case for oceanography. For sure, the alliance-funded oceanographers took over from previous SCOR activities. Yet the collaboration with Spanish oceanographers embarrassed British oceanographers, and more generally the alignment of oceanographic interests with the Cold War agenda irked experts interested in collaborations that crossed Cold War divides. Because of this, British officials became less willing to actively contribute to NATO's scientific affairs, for oceanographic studies had undermined them at Gibraltar and caused frictions in the North Sea. If one adds their proverbial lack of enthusiasm for large investments, then it is not surprising that in the mid-1960s UK officials became less eager to support their US colleagues in the Science Committee.

Furthermore (as the next chapter will show), even the merits in the realm of defense of the scientific program that Robertson had envisioned now came under closer scrutiny. To start with, its chief architect suddenly died, leaving in legacy the need to accommodate a variety of responsibilities in the hands of Rabi, Zuckerman, and other scientists at NATO. Moreover, the principle by which the production of new knowledge best accommodated defense and surveillance priorities was a pillar shaken at its foundation by problematic reviews. A debate now ensued at NATO's headquarters on whether it was the defense scientist, and not the civilian, who should provide advice on defense-related research issues.

And so, during the 1960s, NATO-sponsored scientists continued to pursue novel research, believing that this pursuit was a major contribution to rejuvenating the alliance as a political organization. But that the alliance really benefited from a science program was an assertion now increasingly undermined from within.

THREE

The Storms of Dr. Strangelove: Environmental Warfare and the Science Committee Crisis

The fire that I describe is likely to release the equivalent of 1000 megatons of energy. [. . .] This is the most violent and wide-spread environmental change which can be expected from a nuclear attack.[1]

Seven years after its establishment, the NATO Science Committee entered an existential crisis. Some of its members no longer considered its program as spearheading an innovative set of projects aligned to the alliance's defense and diplomatic ambitions. Others even wondered if time was ripe for shutting it down. What had happened? As this chapter will show, although NATO's research program was in full swing, leading scientists with a stake in the alliance criticized its goals and trajectory. Their criticism paralleled NAC debates on the alliance's strategic posture and major international events such as the 1962 Cuban missile crisis. Because of these conflicts and Cold War ruptures, views in the alliance on the promotion of science markedly changed.

NATO's military authorities debated its future investment in science too, as they looked for a more effective defense strategy. And they reexamined the relationship between scientific advice and the provision (and development) of new weapon systems. In the running of this review, their considered one prospect as particularly enticing: find out how to manipulate, or "weaponize," the environment. This meant, for instance, to look into the characteristics of cyclones in order to find ways to turn them against enemy territories or explode high-altitude nuclear weapons to produce major atmospheric disturbances hitting the enemy's communications. Propagandized by the US nuclear scientist Edward Teller, these far-reaching ideas revealed that the alliance's officials continued to view the natural environment as an item to be made pliable to military operations, but this time not just because of surveillance.

Partly because of Teller's propositions and NATO's revised research agenda, the Science Committee's activities now underwent closer scrutiny and more questions about the effectiveness of its program ensued. Was the committee *really* promoting studies of significance for defense? Or were the scientists that the committee sponsored too interested in fundamental research to offer assistance? And were the committee members too unaware of NATO's key defense requirements to be able to provide advice to its military agencies?

Teller and others now openly questioned NATO's investment in science in general (and fields like meteorology more specifically). This criticism produced the first (and most significant) crisis of the Science Committee in its history. As the crisis peaked, its divisive aspects further convinced the Science Committee members that the promotion of international scientific collaboration on its own was not enough to strengthen the alliance politically or improve the relations between its member states. The negotiations on the committee's future led to a different agenda for NATO science (and the alliance as a whole).

A Slow Start? NATO's Advisory Group on Meteorology

The Science Committee's crisis was rooted in the circumstances of its advisory group on meteorology. Since the committee had agreed that oceanography and meteorology were areas suitable for international scientific collaboration, in 1960 the NATO science adviser Frederick Seitz invited three meteorologists to discuss the establishment of a subcommittee in charge of collaborative research. Yet the invited meteorologists, in contrast with the oceanographers, struggled to elaborate a research program aligned to their patrons' wishes. Within a few years from Seitz's invitation, the polemic on the group's shortcomings escalated, also casting more doubts on the alliance's science program.

When they first met at NATO, the chief research officer at UK's Meteorology Office, Reginald Cockcroft Sutcliffe,[2] Jacques van Mieghem of the Belgian Royal Meteorological Institute, and Arnt Eliassen of the Norwegian Institute of Meteorology, agreed that the alliance could set up a program of meteorological studies. But they were also aware that the space for innovative cooperative actions had considerably shrunk in recent years, since they were already active in a number of other international schemes.

As secretariat members of the World Meteorological Organization (WMO, established in 1951 as an agency of the United Nations), the three meteorologists had contributed to a number of studies showing the potential of international cooperation in providing new insights to global-scale weather phenomena.[3] They also played a leading role in the ICSU-sponsored International Association of Meteorology and Atmospheric Physics (IAMAP).[4] Both WMO and IAMAP had taken responsibility for the organization of IGY meteorology-related projects. Other collaborative endeavors had granted popularity to some of them. For instance Eliassen had taken part in the US Meteorology Project in which, through the use of the mainframe computer ENIAC, the Hungarian-born mathematician John von Neumann had explored for the first time the potential of applying calculating machines to routine weather forecasting work.[5]

The experts that Seitz invited were also members of their own national meteorological offices, and as such were busy looking for ways to improve the provision of weather forecasts at the national level. Meteorological offices had increasingly diversified their clientele to include industrial and agricultural concerns, as well as the military and the state at large.[6] The national weather agencies of Sweden, West Germany, and Britain all had ongoing relationships with the US Weather Bureau and were keen to set up their own numerical weather prediction projects.[7] Presumably Sutcliffe, Eliassen, and van Mieghem also knew that it would have been equally difficult to operate within NATO without treading on others' toes in that the Standing Group had its own meteorology committee.[8]

The extensive activities already typifying meteorology within and outside NATO made these experts cautious. Cooperation on a larger scale in the field of meteorology was a possibility, and some areas were useful for research, including the study of mesoscale phenomena, the structure and dynamics of fronts, the utilization of computers in weather prediction methods, and the study of climate change.[9] But, in contrast with NATO's colleagues devoted to oceanography and radio meteorology, they were assertive about openly tying these subjects to specific defense issues.

Science Committee members might have sensed their impasse from the fact that, after the meteorologists' first meeting, they asked for a "somewhat amplified" and "more specific" plan.[10] Seitz thus called in two more experts to join them: the German meteorologist Hermann Flohn (who had contributed to AGARD's *Polar Atmosphere* symposium) and Daniel F. Rex of the US Chief of Naval Operations office. The newcomers now proposed a study on the genesis of cyclones, for this was a topic with clear implications for defense operations.[11] Cyclones disrupted military operations and communications, and the interaction between jet streams and the morphological features of a territory affected air missions too.

The Science Committee now positively reviewed the abridged meteorology program, but it was not unanimous in recommending that a subcommittee be established. In fact, by the time Seitz's successor William Nierenberg was appointed, the creation of a meteorology subcommittee was still an outstanding item.[12] To an extent, the Science Committee's opposition derived from another issue that had featured in its debates. In the summer of 1959, the NAC instructed a group of prominent committee members under the chairmanship of French engineer Louis Armand to prepare a report on how to increase the effectiveness of Western science. The group, in line with Rabi's and Killian's advocacy, produced several recommendations including the creation of an International Institute of Science and Technology to comprise fields of importance at international level such as oceanography and meteorology.[13] At their meeting of September 1960, however, it was stressed that in the future the committee should recommend only the setting up of "joint working groups of specialists *rather than consultative committees*" (my emphasis)" in the belief that letting specific groups acquire committee status would compromise the effectiveness of the program as a whole by allowing them too much freedom.[14]

The underlying critique was that the science adviser had seconded the DRDs and the US delegation in prioritizing meteorology and oceanography, while the Science Committee members were not at all unanimous in recognizing the need for a defense-oriented science program focusing on these disciplines. Rabi, eager to keep in line with the mixed philosophy advocated by the US delegation, now defended the decisions made the previous year and presented a US statement that recalled the merits of an investment in meteorology and oceanography because of their vital importance for the alliance's defense.[15]

Meanwhile, NATO's meteorologists laid out a program now encapsulating defense-relevant subjects and paid attention especially to a project by a latecomer in the advisory group, the Italian atmospheric physicist Giorgio Fea. A former student of Nobel-winning physicist Enrico Fermi, Fea was recruited in the Italian air force during World War II. At the end

FIGURE 3.1. The expansion chamber that Giorgio Fea designed in the late 1940s. It was installed at the National Institute of Geophysics in Rome and used for artificial rain trials sponsored by the Italian Ministry of Agriculture and the Italian air force. Source: Archivio dell'Istituto Nazionale di Geofisica, Rome.

of the conflict, he was instrumental in setting up the force's meteorological agency and was the first to hold a chair in meteorology at the University of Rome. Fea had also pioneered artificial rain studies in Italy (through cloud seeding) and established a network of meteorological stations covering its territory (see figure 3.1).[16] He was also involved in Europe-wide collaborative exercises uniting meteorological offices in monitoring nuclear fallout from the first nuclear French test of Reggane (Sahara desert) in 1960.[17]

Fea played an important role in NATO's meteorology advisory group, especially by designing a new project on Alpine cyclones in collaboration with Sutcliffe and Vassy. (The latter was drafted to set up a NATO network of ozone-sounding stations.)[18]

The interest in cyclones partly derived from the need to consider major weather events in proximity to launching facilities and air bases. On the occasion of the *Polar Atmosphere* symposium, presentations and discussions had focused on cyclonic and anticyclonic activities at the North Pole, and especially in the Novaya Zemlya region, where one of the Soviet atomic weapon polygons was located. During the symposium, the invited experts—including Sutcliffe and Flohn—discussed the cyclogenesis of polar phenomena as well as very distant climate anomalies associated with cyclones above the polar circle (the so-called teleconnections).[19]

So the advisory group on meteorology promptly agreed to recommend that the *Lee-Cy* (cyclogenesis in the Alps' lee) project be funded, and the scheme received the largest grant among those that the group proposed. Fea and his colleagues thus investigated the jet streams influencing the baric depression over the Genoa gulf, eventually concluding that the interaction of a frontal northwesterly flow with the Alpine barrier was a decisive factor in this cyclogenesis. The study paved the way to the development of forecasting techniques later utilized by the Italian air force's meteorological office.[20] Fea's project was not the only one to have implications for air defense and more projects were now designed, including one on the formation of clouds in the region surrounding Mount Olympus and one by Eliassen on atmospheric waves produced by mountains.[21] On the whole, however, the meteorology program failed to impress Science Committee members. Its shortcomings turn out to be particularly problematic, for in the 1960s the alliance's defense strategy underwent a comprehensive review.

Teller, Zuckerman, and the Prospect of a Nuclear Holocaust

From 1961, NATO had embarked on the important debate on "flexible response," and it was in the aftermath of this debate that key decisions on the future of NATO's science program were made. Advocated by the new US secretary of state, Robert McNamara, the new strategic concept conformed to Kennedy's new vision of NATO's role in the Cold War conflict. In contrast with Dulles's massive retaliation, McNamara's strategy envisaged that using the nuclear deterrent would be the last resort. Although the need for flexibility in planning retaliation had already emerged during the Eisenhower administration, McNamara brought it into the NATO arena after the Berlin crisis.[22] In October 1962 a new international emergency defined by the placing of Soviet nuclear missiles in Cuba further informed McNamara's advocacy of a new strategy, highlighting the impossibility of avoiding nuclear annihilation if a world conflict took place.

Flexible response was being debated at NATO even earlier. In Britain, it emerged as a viable proposition because of Zuckerman's study of the consequences of nuclear bombing.[23] After his appointment as the UK minister of defence's science adviser, Zuckerman had set up the Joint Inter-service Group for the Study of All-Out Warfare (JIGSAW), whose findings were reported to the annual SHAPE conference of defense analysts, "SHAPE-X." Already in 1960, when he learned about SACEUR and US General Norstad's plan for deploying medium-range ballistic missiles (MRBMs) as part of the NATO forces in Europe, Zuckerman vocally objected to the scheme, claiming that "the result would be a devastated Europe covered in radioactive dust." No "acceptable damage" could come from it.[24]

The argument sparked a confrontation with a US nuclear physicist who regularly boasted the merits of nuclear weapons. Edward Teller (like von Kármán and von Neumann, yet another Hungarian-born US scientist) was a key contributor to US atomic weapon programs. Dubbed the "Real Dr. Strangelove" by biographer Peter Goodchild, Teller held questionable views on the use of nuclear weapons as means to address all sorts of problems. He famously sponsored controversial civilian schemes such as the AEC's Project Chariot, in which an artificial harbor would be created with controlled nuclear explosions.[25] By the time SHAPE-X took place, Teller was busy showcasing a newly designed neutron bomb, which allegedly left infrastructure and civilians unharmed while destroying enemy combatants. Zuckerman was unconvinced, however. As he later wrote, the nuclear scientist spoke "only for himself and his band of acolytes."[26]

During his visit to NATO, Teller tried to gain support for US propositions that had already failed to impress Western European allies. Norstad's MRBM proposal had by then met the opposition of French and German diplomats, and SACEUR worried that NATO would face a major diplomatic crisis unless the UK delegation supported him. By then the US administration was about to reject the NATO Secretary General's plan for making available the missile Polaris to France in order to prevent the strengthening of its relations with Germany and the polarization of NATO along competing US-UK and France-Italy-West Germany partnerships.[27]

The UK minister of defense let Zuckerman know that he worried about selling his critique,[28] but the debate on MRBMs continued as Norstad eventually came up with an alternative plan based on tactical (i.e., low-yield) nuclear weapons. This entailed, first and foremost, reshaping the NATO shield using aircraft designed for quicker takeoff and enhanced surveillance capabilities.[29]

But Zuckerman considered SACEUR's new plan "an elegant piece of special pleading" that, however, left open the question of whether escalation could be prevented.[30] McNamara sought to redress NATO's posture more decisively, but Zuckerman's position found many officials, including Rabi, sympathetic.[31] Rabi's colleague Jerry Wiesner, one of Kennedy's closest advisers, dryly remarked that although Zuckerman's views were welcomed by the president's experts, "the Military were likely to take a pretty poor view on his casting of doubts about sacred cows."[32] By 1963, when Norstad was replaced by General Lyman Lemnitzer, NATO was about to set up a multilateral force, partly based on the alliance's own vessels equipped with the newly designed SLBMs.[33]

By then the number of tactical nuclear weapons in Europe had grown threefold (from 2,500 to 7,500) and relations between the United States and partners in Western Europe were now strained. Kennedy mothballed the plans for a NATO nuclear force; something that brought the alliance on the verge of collapse on the occasion of the NAC meeting of December 1961.[34] US allies were even less happy with Kennedy's unilateral decision to remove Turkey's MRBMs in exchange for shipping the missiles in Cuba back to Soviet Russia on the occasion of the Cuban missile crisis. And now it was the US partnership with the United Kingdom to suffer because of this lack of consultation. In 1962, Kennedy's decision to cancel Skybolt (which was promised to the United Kingdom) upset transatlantic relations even more.[35] The only event that made US diplomats less pessimistic about NATO's circumstances was de Gaulle's veto, in 1963, of Britain's entry in the European common market. It showed that the *entente cordiale* was in worse shape than the Special Relationship.[36]

The political debate on NATO's strategic posture affected the discussions on past and future avenues of its science program, also informing its effectiveness as a device of NATO diplomacy. The flexibility advocated by McNamara now put pressure on the alliance's military authorities to review what weapon systems could be innovatively used in warfare activities. And the Standing Group's chairman now solicited AGARD's assistance in forecasting scientific trends affecting defense; its leader von Kármán would chair the new committee (henceforth Von Kármán Committee, or VKC) responsible for the study.[37] The group also included defense research directors from Canada, France, West Germany, Britain, and the United States.

On March 6–25, 1961, one hundred scientists were invited to Allied Forces Southern Europe Command in Naples to contribute to the VKC's first research exercise. The following June the committee submitted an interim report summarizing key findings, and physical environmental sciences featured prominently in the report.[38] One VKC panel of experts had by then underscored their operational significance: "since they pertain to environment, [they] are important for all military missions."[39] Meteorology and oceanography were singled out recognizing that some phenomena, including "density variations, internal waves, deep currents and turbulence," were poorly known. In terms of forecast, the experts remarked that by 1975 more satellites could provide global observation of clouds and thus revolutionize military operations. They also hinted, rather vaguely, at the possibility of manipulating the weather.[40]

The search for more flexible weapon systems prompted the committee to further examine this hazily defined topic. The VKC instructed a second round of consultations, which took place on June 12–30, 1961, at NATO headquarters in Paris. The invited experts now considered a wider range of new weapons, such as those attacking "basic ecological dependencies," through chemical and biological weapons and ways to control the climate.[41] The list of weapons featured heavy radioactive fallout to affect urban populations and tampering with a country's agricultural supply by disrupting its climate.[42] Diverting sea currents and generating cyclones (the topic of NATO's most expensive meteorology project) were also considered. As Jacob Hamblin has recently remarked, at this point in history environmental warfare was "unconventional," literally, for it was not even classed as an act of war and no international convention forbade it.[43]

The experts invited by NATO also argued that nuclear weapons could be used for the purpose of environmental modification (producing fallout, blackouts, and contamination by radiation). They also extended these considerations to the use of chemical and biological weapons, envisaging the possibility of attacking ecological communities and affecting "interdependent living things."[44] John von Neumann's early studies on climate manipulation had a tremendous impact on these reflections. In an article published in the magazine *Fortune*, he claimed that weather control would be used regularly in the future to improve agriculture and commerce. Although he did not single out military uses of climate, the experts invited by the VKC panel quickly reframed von Neumann's analysis to emphasize them.[45] Viruses, molds, rusts, and blights could contaminate foreign environments. Attacking the vegetation with chemicals could inhibit guerrilla warfare (as the conflict in Vietnam would soon show).[46]

VKC members now agreed to appoint a panel to look further into the potential of environmental warfare, and USAF Colonel Pharo A. Gagge, a biophysicist who had been responsible for studies of pilot performance in high-altitude flights and had recently worked in a DoD weather-modification unit, agreed to chair it.[47]

Controversially, Gagge also agreed to invite the atomic weapons enthusiast Edward Teller to take part in the exercise. In this way he reignited the early antagonism with Zuckerman, for "Sir Solly" represented Britain in the VKC. From Zuckerman's perspective, it was just the wrong moment. Robertson, the ex-SHAPE science adviser who had shown the virtue of blending together different ideas and approaches at NATO, had just died, killed in a tragic car accident in Pasadena on August 26, 1961. His death deprived Zuckerman of a precious friend and ally. Now the frictions between Teller and Zuckerman could no longer be contained.

The VKC environmental warfare panel met from October 30 to November 9, 1962. Teller, who featured as special (panel) adviser, opened proceedings with a paper, "Nuclear Explosions," that admittedly did not beat around the bush. He enthusiastically argued for the merits of high-altitude nuclear blasts that could generate winds and hurricanes and could disrupt enemy defenses and communications.[48] The nuclear test *Starfish Prime*, a 1.4-megaton device exploded four hundred kilometers above the Pacific Ocean, had recently provided evidence on the potential of high-altitude nuclear explosions. To his peers at NATO headquarters, Teller enthusiastically confirmed that the test had been a great success; it disrupted satellite operations and affected the earth's magnetic field. An auroralike glow above the Pacific sky heralded the achievement.[49]

Teller's opening lecture thrilled the participants. The panelists went on to examine the merits of environmental warfare from outer space onto the earth. Oceanographers explained that tactical nuclear weapons could generate tidal waves and divert ocean currents.[50] The genesis of cyclones also was discussed at length because it lent support to plans to manufacture tornadoes. The panel's final report concluded that "if a NATO group of meteorologists undertakes an analysis of atmospheric behavior" in the next two to four years, "the first weather modification experiments" could be designed.[51] As James Rodger Fleming and Kristine Harper have shown, this enthusiasm for weather modification was by and large misplaced. By then, the US military had already sponsored trials that had gone dangerously wrong, convincing the military authorities to quickly cancel projects and hush up the results.[52]

When the VKC members reviewed the panel's report, their reactions were mixed. Zuckerman openly attacked the conclusions reached by the panel as "highly speculative" because the dispersal of cloud and fog rested on little experimental work and cloud seeding was unpredictable. The frustrated Zuckerman also remarked that the panelists had used "a queer rag-bag of meteorological terms."[53] The US deputy assistant secretary of defense for research and engineering, John B. Macauley, who had recently replaced Robertson in the VKC, was a little more hopeful because the panel was led by a US defense scientist.[54]

A supplementary report was prepared by the VKC and released in September 1963. The section on environmental warfare focused mainly on high-altitude nuclear explosions and their effects on the ionosphere, judging this to be the only promising environmental weapon for future development.[55] The report thus doomed attempts to sponsor research in the area in the foreseeable future.[56] Many at NATO took notice. But so did the nuclear weapons champion who had stirred the debate on environmental warfare.

Storming Dr. Strangelove

Teller and Gagge, possibly because of resentment about Zuckerman's comments on their report, now stirred a controversy on NATO meteorology. On January 7, 1963, they had already sent a letter to von Kármán stressing that meteorological research at NATO had not developed enough. Todor promptly forwarded the letter to the newly appointed NATO science adviser, MIT physicist William Allis.

When on January 28, 1963, the members of the advisory group on meteorology met, they cast aside Teller's criticism, reiterating that meteorology had important defense implications but also stating that it was incorrect to suggest that NATO had not done enough.[57] At this stage, the group succeeded in downplaying Teller's remarks partly because the issue was bundled with another item on the Science Committee, namely a (failed) plan to establish a NATO meteorology center promoted by US physicist Fred Singer. Furthermore, by then a proposal by MIT President James Killian for a NATO international institute of science and technology also was being discussed.[58]

The polemic on the military uses of meteorology at NATO did not fade away though. Teller had made a similar proposal to sponsor weather manipulation research at the US National Academy of Sciences and his proposition had encountered opposition. Therefore he felt a need to set the record straight with his opponents on both sides of the Atlantic.[59] In September 1963 the real Dr. Strangelove sent another letter to NATO delegations stating that the alliance should use its ships and aircraft for the purpose of carrying out more meteorological observations. By making use of sondes to acquire new data, he argued, NATO could produce more numerically accurate weather prediction methods. These, he naively estimated, would support projects to manipulate weather.[60]

Teller's propositions reached the Science Committee again. And this time the Standing Group representative indicated that his organization was very interested in Teller's plans, thus exacerbating the dispute. One committee member now urged the meteorology advisory group to turn its attention toward "problems of military significance."[61] This time, it was impossible for its members to take no notice of the disapproving remarks. When they met in January 1964, the meteorologists pointed out that the scientific tasks Teller advocated could not be carried out because NATO commanders "do not usually control the movements of vessels during peacetime."[62] The minutes of the meeting also suggest a dust-up with the Standing Group representative who attended the meeting. The abridged version of the minutes records a "lengthy [unminuted] discussion," at the end of which van Mieghem expressed his gratitude to the Standing Group representative for "presenting his [unminuted] views."[63]

Now the conflict brewing over the trajectory of NATO meteorology informed Science Committee proceedings. When in August 1964 its members met to review NATO's meteorology efforts, Allis informed the other members that Teller appeared to be satisfied with the remarks that the advisory group had made. Science Committee members sympathetic to his position, however, inquired whether defense research should still be considered within the group's terms of reference.[64] Thus, in March 1965, Sutcliffe voiced concerns about the range of activities that could be carried out, something he referred to with "a degree of uneasiness."[65] The meteorological advisory group had attempted to find a useful function notwithstanding the background of intense meteorological activities internationally and within NATO. Sutcliffe also recalled that the group had never been given committee status.

Sutcliffe's words ignited a dormant conflict on the trajectory of NATO's science program. The Norwegian representative Svein Rosseland argued that the alliance should sponsor only defense-oriented research and proposed that the advisory group be dissolved. Rabi argued against it, though, keeping in line with the mixed philosophy that had been at the core of the program. And for once Louis Néel agreed with Rabi while urging the meteorologists to do more.[66] Teller's storming thus did not produce the changes that he had advocated. NATO would not be sponsoring environmental warfare studies—at least as part of its research program. But his intervention in NATO's scientific affairs was to prove toxic.

The Science Committee at a Watershed

Teller's reproach seemed to demonstrate that a gap existed between what the committee aimed to promote and what NATO defense authorities needed. Rabi's efforts to promote research embedding a defense agenda now appeared less than successful. Robertson had been the only one to recall the importance of one for the other, exporting a science policy model that had contributed to his popularity at home.[67] But he was no longer alive. His correspondence suggests that he was preparing to bring his influence to bear on the ongoing NATO debate on science. In March 1960 he wrote to his colleague Warren Weaver, urging him to counter opposition to projects outside the area of basic research. Pure science provided the answer to fundamental defense issues, he contended.[68]

NATO's science program was reassessed in Robertson's absence. The controversy over meteorology proved particularly noxious because of a parallel review on providing scientific advice to military authorities. The DRDs carried out the preparatory work, although its members had been cautious about making deliberations because of the ongoing quarrel over AGARD's role in the alliance, an argument still simmering.[69]

Meanwhile Rabi realized that the US-UK partnership that had fortified Science Committee activities was now dissolving. Early UK complaints about financial commitment had now become raucous objections. Already in 1961 Nierenberg had informed Rabi that Zuckerman could not rally sufficient enthusiasm in Britain on Killian's proposal for a NATO international institute of science and technology; a proposal that had already featured in the report prepared by the Armand group.[70] Nierenberg later wrote him that knotty issues existed: "the major question being Great Britain, and here we depend on yourself, Killian, Cockcroft, Sir Solly, the American Foreign Office, Mr. Stikker and God."[71] Zuckerman now admitted to Rabi that his government viewed the NATO science program as an expensive, and not entirely useful, toy.[72] Rumors about the possibility that a new committee would take charge of defense research reached Rabi's ears and the possibility that Zuckerman could play a role in it troubled him.[73]

On March 22, 1963, the DRDs concluded that the transmission of scientific advice between the Science Committee and national military authorities had not been adequate.[74] They thus appointed a four-man exploratory group chaired by the electrical engineer (and France's DRD) Pierre Aigrain. Acquainted with the former NATO science adviser Frederick Seitz and well connected with the administrators of US and French armed forces, Aigrain took over a task meant to help resolve differences between NATO's science and technology groups, with the assistance of the representatives from Italy, Britain, and the United States.

The conclusions that Aigrain's group reached suggest that by now French and UK officials were prepared to join forces, isolating Rabi, and seeking to persuade the new representative in the US Department of Defense that NATO should start a defense-driven research program. The report that the group completed on October 17, 1963, discussed at length the current standing of NATO science. But its conclusions ruled out greater influence in the alliance for AGARD.[75]

The report was equally critical of the Science Committee, though. The committee had been unable to stimulate military-related science, partly because of the nonmilitary training of the science adviser and the other committee members.[76] Aigrain's group concluded that it had been wrong to assume that "by simply adding a civilian science adviser and a scientific committee to the existing NATO machinery, flow of scientific advice to the civil and military authorities would automatically be forthcoming."[77] It is worth recalling at this point that the controversy over meteorology had unfolded in parallel and Teller's second letter was sent weeks before the report was completed. The case was a perfect illustration of what Aigrain's group sought to demonstrate. The exploratory group concluded that the DRDs should no longer operate as an informal forum but needed to become a working NATO committee. Though hoping that the Science Committee would "retain its right to make observation on military points if it wishes," the responsibility for offering sponsorship to defense-related work should now be taken over by the DRDs.[78]

When the defense research directors met again on November 21, 1963, they agreed to appoint Zuckerman as the DRD Committee's temporary chairman, waiting for the NAC to approve the suggested changes to NATO structure. Zuckerman rejected considering "civil" and "military" science as separate corpuses of knowledge, but he agreed on the need for an organization capable of encouraging military science. It had been inappropriate for the Science Committee to be burdened with military questions for which its members had neither competence nor security clearance.[79] French, Canadian, Dutch, and Norwegian delegates (but, notably, not the US representative) endorsed Zuckerman's view.

Why had Zuckerman's position shifted so dramatically? His initial support for Rabi's (and Robertson's) early propositions faded partly because of the tensions that typified transatlantic relations in the talks that followed the Cuban missile crisis, thus further recalling, as the present study has so far shown, that the planning of science actions and policy, and NATO diplomacy, went hand in hand. It is also true, however, that, especially after the failure to complete the Blue Streak project, Zuckerman realized that major differences existed in the shaping of the science-military relationships in the United States and Western Europe.

Zuckerman had been privy to the members of the "visible college" forging leftist stances on UK science policy in the interwar years and advocating science planning on a socialist model, but he had eventually ditched the rhetoric that was distinctive of colleagues like John Desmond Bernal.[80] Yet he was now becoming equally disillusioned with those ideologues of scientific freedom that disregarded the negative impacts of the free-wheeling and fast-growing US military-industrial complex on the development of other countries. Already in 1960 he had written to Seitz urging him to consider that Western Europe did not possess a military-industrial complex comparable to that of the United States.[81] His disappointment added up to his complex role as Ministry of Defence science adviser, as he clashed frequently with the UK minister of defence and, in October 1964, when Harold Wilson formed a new Labour government, he agreed to become the government's chief scientific adviser.[82]

Néel was also in a difficult position and, in contrast with his predecessor Danjon, he now agreed with Zuckerman about the existence of key differences between the research communities of the United States and Western Europe. The problem was not defense research as such; both Néel and Zuckerman wanted more of it at NATO. But they disagreed on the convergence of defense and civilian research tasks in one forum. Néel was by now busy promoting alternative collaborative frameworks, such as, for instance, the Franco-German project for the construction of a high-flux research reactor in Grenoble.[83]

"Doom and gloom" existed not only in Zuckerman's and Néel's camps. US officials agreed that the NATO science program had not delivered enough. On March 4, 1964, Nierenberg completed a report in which he claimed that the program had performed well but had little chance of improving further. The scheme on research grants was the least successful. Oceanographers had been responsive, but "an attempt in the meteorological sciences did not fare as well." Collaborative research had proven difficult; "money in itself is not the total answer."[84]

After the creation of the Committee of Defence Research Directors (DRDC), the DoD deputy director of defense research and engineering, John L. McLucas, was appointed the new NATO science adviser (and DRDC chairman). For the outgoing science adviser Allis, the event represented a watershed for the Science Committee in that the committee now "will have to depend upon itself to think of new ideas for things to do."[85]

Zuckerman's Resignation and Rabi's Scheming (While Dr. Strangelove Is Back)

The arrival of McLucas at NATO promised to shake up its research establishment. He did not wish to make the alliance a champion of science diplomacy and contended that from then on NATO should focus on defense research. On July 31, 1964, he informed journalists at NATO headquarters that the alliance would not "evolve into the political, cultural and economic union that its earliest proponents had hoped." NATO "will fundamentally remain a defense military alliance," he stated. The alliance's problems were essentially military and "NATO doesn't need more scientists. [. . .] It needs more application of science to the military problems of the day."[86] In contrast with all his predecessors, McLucas was employed not in academia but in the DoD research and engineering section.[87]

Under McLucas's chairmanship, the Science Committee slowly plunged into a state of crisis. From 1964, the committee's budget no longer increased and stagnated for the following three years, when the NAC approved a marginal increase of 5 percent. Moreover, while funding for science fellowships and ASIs was at the same level as that of previous years, in 1964 the research grants budget marginally shrank and remained the same for the following two years. Finally, from 1965, the Science Committee budget was no longer funded according to the conventional burden-sharing formula and a new system was introduced.[88] This meant that the US share decreased and, by 1966, the NATO science budget was mainly subsidized by Italy ($390k), followed by Britain ($364k) and West Germany ($345k). France and the United States would now share the burden in equal amounts ($312k).[89]

British and French delegates now reunited in campaigning against the US proposal for a 5 percent increase in the budget to match rising inflation. Their opposition was particularly troubling because it echoed existing diplomatic tensions. The French opposition jibed with its officials' resentment for the role that NATO allies had played in the settlement of the Franco-Algerian conflict. In the events leading to the Evian peace agreement with the Algerian rebels, several French officials suspected both Italian and US governments had secretly supported the Algerian cause.[90]

The suspicion paralleled the conflict between US, French, and British administrations on the sharing of NATO's nuclear deterrent. Although a number of alternatives had been considered by Kennedy's and Johnson's administrations, including the possibility of establishing a joint nuclear force, in the end they ruled it out, keeping the British (and French) out of NATO's "nuclear business."[91] The British delegation continued to be assertive toward the other allies partly in the recognition that it was, aside from France, the only other nuclear power in the alliance and thus prioritizing bilateral relations with the US to multilateral coordination.[92] Furthermore, the Pâques affair poured oil on the already inflammable French relations with NATO partners: in August 1963, George Pâques, a French member of NATO's international staff, was found guilty of having passed restricted information to Soviet agents, thereby making available details on recent plans to develop psychological warfare research.[93]

While the British delegation at the Science Committee did not want to be perceived in US eyes as supporting the French attitude, it essentially shared it. The UK Foreign Office thus openly "encourage[d] the French to stand firm"[94] in their objection to raising the budget, although some British civil servants more belatedly warned that "to ally ourselves with the French in cutting down a NATO activity in present circumstances seems a dubious proposition politically."[95] Unsigned Foreign Office notes provide useful insight into this stance through their abrasive wording: the cuts to the Science Committee's budget represented a "good opportunity to do a little bit of streamlining" while the United Kingdom should support "military science" only (i.e., the DRDC).[96]

Zuckerman did little to redress this trend. So while continuing to be the UK delegate at the Science Committee, he did not attend meetings and was often replaced by the University of Cambridge professor of metallurgy Alan Cottrell. Zuckerman had been busy reviewing the relationship between science and military affairs, putting forward controversial theses in an edited collection summarizing the prestigious Lees Knowles lectures delivered at Trinity College, Cambridge. Using the case of the canceled British aircraft project TSR-2 and the missile Blue Streak, he recalled that the growing investments in defense research projects did not necessarily increase the security of nations. He thus advocated a redeployment of technological resources and reiterated his stance on nuclear exchange, now more strongly advocating arms control. His conclusions were virtually incompatible with his role as MoD science adviser: "the insistent call made by the military machine upon the research and development resources of industrialized countries holds back progress."[97]

Representing Britain at the Science Committee had clearly become a burden for Zuckerman. On the contrary, Cottrell was happy to quarrel with Rabi, especially by objecting to his efforts to take the Science Committee out of the quagmire. In February 1965, Rabi offered a proposal for a high-level public conference on the partnership between science and the military in providing for the security of the North Atlantic community, but Cottrell, McLucas, and Néel objected. When the discussion turned toward a Danish initiative for visiting fellowships, Rabi pounced on the issue again, asking whether military science was to be included in the scheme. Cottrell urged him "not to over-emphasize" it.[98]

Rabi had to face considerable opposition in his efforts to revamp the Science Committee. McLucas's former head, the US assistant secretary of defense for research engineering, Harold Brown, was not enthusiastic about the conference either and believed it "a patent attempt to reawaken the interest of the Council in the Science Committee."[99] Don(ald) Hornig, the science adviser to the new US President, Lyndon Johnson, was also skeptical.[100] So was the newly appointed NATO Secretary General, Manlio Brosio. Formerly Italy's ambassador in Washington, DC, Brosio believed that the committee "should redefine his own position and proceed to develop other aspects of the programs."[101]

Rabi now hoped that Zuckerman could assist him, thus restoring the partnership that had once made the Science Committee thrive. In May 1965, he urged Zuckerman to attend the forthcoming committee meeting. Crucially, Rabi reminded him of its importance for the alliance's *political* stability, in that it was the only "functioning organ to which the French are contributing" and stated his worry about Brosio's attitude and the difficulties now arising with McLucas and the DoD. Views on support for fundamental research had dramatically changed and so had relations between the DoD and the State Department.[102]

Zuckerman eventually sought to help his US colleague, but he disagreed with Rabi about big projects funded through NATO grants as a way out of the committee's crisis. The British zoologist now proposed that the committee concentrate on the bourgeoning area of human factors (for which a NATO subgroup existed) and invest in behavioral and social sciences.[103] On March 23, 1965, he informed Rabi that his plans for a conference would be discussed at the DRDC meeting. Zuckerman was not against Rabi's proposal, but he was hardly thrilled by it.[104] Furthermore, DRDC's activities kept Zuckerman busy as the new committee was proving far more difficult to run than expected. When, on October 8, 1964, the DRDC met for the first time, early discussions on weapons procurement were not conducive to successful collaboration.[105]

With Zuckerman now anxious about the DRDC and Rabi searching for ways to revamp the Science Committee, the figure they both despised the most suddenly reappeared on the NATO scene. In August 1965 the egocentric Teller returned to Europe to feature in a number of meetings, boasting new plans for cooperation: the alliance should now embark on a space mission to avoid further tarnishing its defense research agenda.[106] McLucas shared Teller's views (in fact, he was the main sponsor of his visit). Rabi and Zuckerman did not comment although they shared contempt for Teller's initiatives. Zuckerman wrote to "Raab" years later that "either people like you and I belong to an irrational world, or he does. [. . .] His malignancy may be encouraged by the vested interests which are opposed to a peaceful world [. . .] as a kind of Dr. Strangelove he appeals to the illiterate pressmen."[107]

McLucas's well-orchestrated visit of Dr. Strangelove did little to redress NATO's coordination problems. The DRDC met once more in 1964, and then Harold Brown sent apologies for his absence at the forthcoming meeting of October 1965 in a letter that was like a death sentence for the recently established committee. The previous two DRDC meetings had been "interesting," but the fruits of its first year "had been less than we had hoped for."[108] He concluded that the time had come to "develop a more effective modus operandi."[109] At that meeting, he raised again the question of the committee's future, and Zuckerman pointed out that the problem was whether the DRDs could "practically and collectively deal with the question of cooperation in R&D."[110] On March 29, 1966, the DRDC approved a document formalizing its dissolution just one year after its establishment.

The NATO crisis of 1966 further refined the process of change. On the very same day the DRDC approved the dissolution document, de Gaulle instructed that France leave NATO's Allied Command Europe. He also informed the alliance's secretary general that France would not join NATO's Nuclear Planning Group and asked for the removal of the alliance's headquarters from Paris.[111] While in principle de Gaulle's decisions complicated the management of NATO affairs, they actually allowed swifter approval of changes already being discussed at NAC level. Headquarters moved to Brussels. And in September 1966 procurement responsibilities passed on to NATO's Naval, Air Force, and Army Armaments Groups. A new Defence Research Group (DRG) took responsibility for defense research. Its relations with procurement, however, would be much looser than that previously assigned to the DRDC, thus preventing the problems that had led to the dissolution of its predecessor.

Rabi now hoped that, once Zuckerman's perils with NATO's defense research had ended, Zuckerman might be persuaded to return to the administration of the alliance's science program. Shortly after the DRDC dissolved, Rabi informed him that he wanted to discuss the fate of the Science Committee; it was "essential that you and I, two old hands in the game, should get together."[112] He was thus extremely disappointed when Zuckerman finally resigned because it "is proving impossible for me to give my personal attention to the affairs of the Committee." Rabi replied with bitterness that he was "disturbed that you are compelled to resign."[113]

The end of the 1960s was a time of change in the sciences in general, and studies on the natural environment more specifically. Although the visionary, but possibly quixotic, Teller continued to view the environment as something that science could make pliable to the defense of the West, Rabi hoped to use it in the alliance's diplomacy realm. He, more than anyone else, understood that NATO should undergo a process of political transformation in order to find new strength. Such rejuvenation depended on proposing something new and different.

Alert to the rebellion then mounting in Western societies (in schools, universities, and factories), he understood that an inward-looking NATO framing itself exclusively in the defense arena would be struggling for survival. From its inception, Rabi had always perceived the defense of the West as one compelling convincing military strategies as much as efforts to unite through diplomacy and to win "hearts and minds." As the next chapter will show, the environment was at that point mobilized at the US State Department not because of its defense implications, but because of its power to rejuvenate the alliance as a political organization; to promote its "greening." The respected and celebrated US scientist Rabi now foresaw, to use a figure of speech, that wearing the emerald green used by the environmentalists over the traditional khaki-colored uniform might rescue the beleaguered North Atlantic Treaty Organization.

FOUR

Launching NATO's Environmental Program

In the midst of great advances in living standards we have at the same time an increasingly impoverished environment which is an almost inevitable consequence of the heedless application of science and technology.[1]

In the 1970s NATO embraced a "third dimension" beyond its political and military ones, seeking to promote actions addressing the problems of advanced societies, and especially environmental degradation. According to historian J. Brooks Flippen, the new dimension heralded the beginning of a period in contemporary history characterized by environmental diplomacy, that is to say, "a new phase in international relations that promised collective action for global sustainability."[2] The term is associated with the political trajectory of US President Richard Nixon, since he is one of the first world leaders deemed to have championed environmentalism internationally.

But the reverse of Flippen's definition is equally applicable to NATO's transition. Nixon's environmental diplomacy instigated a dialogue on global sustainability in an effort to redress, or even *evade*, thorny issues marring the cohesiveness of the alliance. France's exit from the NATO-coordinated military structure, the mounting wave of dissent for US involvement in Vietnam, and the search for political unity within the alliance (including the problems with its Science Committee) convinced the US president to think innovatively about NATO's circumstances.

The result was enigmatic in any case, which is why several scholars have recently devoted attention to Nixon's initiative and its main outcome at NATO; the newly established Committee on the Challenges of Modern Society (CCMS), in charge for the alliance's environmental program. Hamblin has convincingly shown that US allies in Western Europe initially resisted Nixon's attempt to take the environment in NATO's ambit, especially when the alliance's intervention prevented extending the dialogue on sustainability globally to non-NATO countries.[3] And Macekura has recalled that Nixon's initiative was meant, moreover, to stave off the debate on economic assistance to developing countries.[4]

It is important therefore to look more closely at NATO's own political debate, and especially the circumstances of the Science Committee, to explain the hidden agenda behind the alliance's "greening." We have seen that by 1966 Rabi understood that a change in the alliance's funding philosophy was needed in order to save the committee from dissolution and retain the diplomatic benefits that he believed would derive from transatlantic scientific collaboration. And so, in the context of committee meetings and especially on the occasion of the 1968 celebrations for the tenth anniversary of its foundation, Rabi sought to persuade his peers that environmental degradation was a far more compelling issue, one deserving NATO's attention.

Yet Rabi's initial proposition was meant to evade equally significant issues including the diplomatic tensions that had recently led to the crises of the alliance and its Science Committee and the mounting criticism, among allies, of its political and strategic agenda. In this respect, Rabi's proposal manifested for the first time the US administration's ambition to use the nascent dialogue on environmental degradation as a diplomatic device. Within a few years of Rabi's speech, NATO's traditional science diplomacy would thus morph into environmental diplomacy, also paving the way to Nixon's intervention, the creation of the CCMS, and the arrival at NATO of a new generation of diplomats versed in environmental problems, such as Russell Train and Eduard Pestel.

A Technology Gap?

By 1967, the debate on the Science Committee's future further deepened its crisis, especially because a new element made these discussions more problematic. The group of delegates from NATO's less developed countries (LDCs, i.e. Italy, Greece, Portugal, and Turkey)[5] had recently been more vocal in calling for a major restructuring of the committee so as to take into greater consideration how science and technology could propel economic growth.

The controversy emerged from an evaluation of what the committee had achieved. The United States was on the brink of a second industrial revolution, thanks to the defense-driven stimulus to science and technology. But the LDCs had struggled to develop an independent research and development sector. The investment had been particularly patchy because the funding lavished on defense research had produced dependency on US provisions and had not sufficiently stimulated an autochthonous research complex. NATO support had done little to redress this trend. And in the case of Greece, economic troubles combined with a political crisis led to a putsch in May 1967 (bizarrely originating in a NATO plan against insurgency).[6]

By then, other initiatives had pushed the alliance's political debate away from US foreign policy objectives. The Germans had been more vocal about the need for better relations with the Eastern Bloc. And in that same year, the Belgian foreign minister Pierre Harmel filed a report calling for NATO to be the harbinger of more peaceful relations between countries and not just a defense alliance.[7]

The LDCs' representatives on the Science Committee argued that beause the decade-long NATO investment had not yielded enough, the committee should introduce new sponsorship schemes. The Italian Amedeo Giacomini and the Portuguese Carlos Alves Martins had been particularly vocal in stating the need for greater emphasis on economic development. As head of the National Institute of Electroacoustics, Giacomini had been involved in NATO's scientific program from its inception as Giordani's alternate. He was now familiar with the ways in which the committee worked and, in contrast with colleagues from Britain and the United States who were supported at home, he single-handedly managed committee business and reported at regular intervals to the National Research Council's head.[8]

The discord stirred up by the Italian and Portuguese delegates typified the proceedings of the Exploratory Group of Six, which was set up in 1966 to address the Science Committee's shortcomings. The group included Rabi, Néel, Alan Cottrell (who was called in to replace Zuckerman), West Germany's new delegate Eugene Pestel, Giacomini, and Martins. In October 1966, a tense Science Committee meeting took place in Portugal. In open and restricted sessions, the frank exchanges between delegates further compromised the search for viable solutions. The meeting started on the wrong foot, as Néel, to keep in line with the recent UK-French convergence on keeping NATO's funding for science in a stationary state, announced reservations about the budget proposed for the following year.[9] The quarrel continued at the next meeting of the exploratory group. Giacomini now requested that the group encourage the adoption of a funding philosophy more aligned with the original Three Wise Men's design for NATO's science initiatives.

When such a proposal did not find sufficient support within the group, the NATO science adviser, McLucas, agreed to have another discussion with LDCs' representatives. He eventually confirmed his support for a separate program targeting underdevelopment.[10] At this point, the discussion reached a deadlock because such support would have required increasing the committee budget, which was exactly what British and French delegations disagreed with. So in January 1967, the six met again only to provocatively assert that time was up for the committee's dissolution. Little could be done to implement its plans without a new allocation bearing a more realistic relationship to the magnitude of NATO's science program.[11] They also asked for the appointment of a science adviser since the Secretary General Brosio had recently appointed the German Rudi Joachim Schall in an acting role wanting the committee to be chaired by a low-level official.[12] This signaled an effort to disengage and cut back.

Alves Martins and Giacomini now prepared a memorandum outlining their requests: a quota system for the fellowships, and a new allocation for LDC-led projects of no less than a third of the funds available in the research grants scheme.[13] But Rabi opposed the plan since these proposals entailed using NATO funds as aid and the US State Department was against using the sponsorship of science projects as economic assistance.[14] The opposition drew on another recent proposition that had attracted the NAC attention and that the director of State Department's International Scientific and Technological Affairs Division, Herman Pollack, had instructed Rabi to categorically reject. In the spring of 1966 the Italian Minister of Foreign Affairs, Amintore Fanfani, presented what Pollack recalled as a "somewhat vague and generalized statement" concerning the existence of a technology gap between US and Western Europe.[15] The Italian politician had also put before the NAC a proposal for financial aid on a variety of research items including atomic energy, computers, space research, satellite rockets and aeronautics.

Behind the initiative was one of Italy's negotiators on nuclear energy, Achille Albonetti.[16] Fanfani and Albonetti believed that US economic assistance to science and technology could have restored balance in the production of new knowledge and wealth between the two sides of the Atlantic. Fanfani reiterated his proposition at the OECD meeting of November 1966 and the NAC discussed it again on March 1, 1967. The Italian initiative failed to impress US officials, though; at one point it was alarmingly dubbed as a "Marshall Plan for science and technology."[17] Donald Hornig also confirmed that the technology gap was a European problem; there was "little that the US government can or should do by way of direct assistance."[18]

Pollack also requested that statistics be made available on the alleged gap. The study of US patents in Western Europe, the technological balance of payments, and research and development expenditure thus showed no evidence of disparity. In some areas (shipbuilding, railways, nuclear power plants, chemicals) Europe was "at least as good as the U.S.A."[19]

Pollack now urged Rabi to adopt an encouraging attitude with Giacomini while avoiding any firm commitment.[20] Rejecting outright Fanfani's proposal meant compromising diplomatic relations with a trusted ally in a moment when key partners like France and Britain had been difficult. Moreover, by 1966 Italy had become the largest investor in the Science Committee and its economy was thriving. And the Italian's current posture, which emphasized openness with Arab countries in contrast with the rigidity of crumbling imperial powers, was welcomed in the United States.[21] Italy's covert support to Algerian independence, in a time when the US administration had also taken a conciliatory stance, and the search for a diplomatic solution to the Arab-Israeli war, represented the cornerstone of this new approach.[22] Its accession to the NATO Nuclear Planning Group (established in December 1966 to ratify the alliance's nuclear strategy and comprising the United Kingdom, the United States, and West Germany) further confirmed its newly acquired status.[23]

In April 1967, Rabi considered how to align Fanfani's proposal to US interests. The possibility existed that his country's research agencies might establish new collaborations with European ones in some key areas. The IBM director of research, for instance, had already manifested an interest in establishing a computing research center in Western Europe.[24] Meanwhile, Giacomini hosted Alves Martins and committee delegates from Greece (M. Angelopoulos) and Turkey (Nimet Özdas) in Rome and, after their meeting, they agreed to leave the committee if their proposals were rejected again.[25] Rabi now sought to mollify Giacomini by supporting his plans for siting a new European computer center in Italy. Two more propositions (one for building an oceanographic floating laboratory) further helped Rabi to pour cold water on far-reaching propositions elaborated by the LDC representatives.[26]

But Rabi's negotiations did not end the committee's crisis. The US science ambassador was particularly anxious about Brosio's reluctance to appoint a new science adviser. In December 1967, he confirmed to McLucas that the committee's survival was decisive for the alliance's future because it now represented a key diplomatic device strengthening it: "I did not see Mr. Brosio. [. . .] I also despair about the Science Committee. There seems to be very little possibility of making NATO understand the importance of holding on to one thread of sanity and emergent energy—the whole ball of wax which is NATO." He teased, "perhaps when you come there [to NATO headquarters] we can both go to Brosio and see if we can fire him with zeal."[27]

Brosio's resistance presumably drew on his resentment for the US rebuttal of Fanfani's proposal.[28] He also foresaw that the Science Committee would be playing a somewhat less relevant role after the DRDC establishment. Rabi never sent the nasty reply he had drafted after receiving Brosio's apathetic note, but he asked to meet the secretary general at his next visit to Brussels.[29] The *tête-à-tête* did not help. Days before the committee meeting in February 1968, the two were bickering: "You should not expect us to submit a new program the day after tomorrow," stated one. "No science adviser will be nominated any time soon," replied the other.[30] So, as McLucas let Killian know, the tenth anniversary of the committee's foundation was now forthcoming but there was no festive mood.[31]

In January 1968, a US State Department document summarized the problems with the committee and remarked, worryingly, that its mission had partly failed.[32] The document contended, however, that there was room for hope, especially if Rabi could "take quick advantage of opportunities for constructive action presented by scientifically interesting events."[33] Environmental disasters were now mentioned in passing as they had recently occupied some space in the media; and one in particular, the oil-spill incident of the *SS Torrey Canyon* tanker, was cited as an example. The event acted as a catalyst for planning committee activities ahead. Rabi's handling of the disaster, actually, helped to advance the US State Department's plan to drain the swamp now restraining the committee.

SS Torrey Canyon and the Tenth Anniversary of the Science Committee

On February 19, 1967, the oil tanker *SS Torrey Canyon* left a refinery in Kuwait with a cargo of ninety thousand tons of crude oil. The ship reached British waters and on March 18 ran aground on the Seven Stones reef between Cornwall and the Isles of Scilly. The oil it carried poured into the sea, creating the largest spill up to that date.

The UK government's science adviser, Solly Zuckerman, was called in to lead a military-style relief operation of NATO proportions.[34] He first decided ("regardless of salvage operations and legal niceties") to place aluminized bombs above the oil in each tank. But the plan had to be aborted when the ship broke up in two. Zuckerman thus instructed crews to set fire to the oil, and RAF bombers used napalm and chlorate bombs over the slick.[35] But it soon appeared that while the explosive was effective in burning some of it, it also made the remaining pollutant mix with water. By March 30 the firing proved ineffective.

Operation Mop-Up was, of course, the subject of intense media scrutiny, and the British government wasted a quarter million pounds sterling per week in cleaning the Cornish coast.[36] The *Guardian* correspondent alleged that the amount of detergent had caused more harm to marine life than the oil slick. Zuckerman rebuffed the accusation. The journalist had got the calculations wrong: "unless people go on pouring the stuff needlessly on some beaches, we no longer expect anything significant."[37] But the handling of the *SS Torrey Canyon* disaster damaged his reputation, especially after the report of the UK House of Commons Select Committee on science and technology recalled lack of coordination between ministries.[38] Zuckerman thus wrote to the prime minister that the report "has been slanted by selecting passages out of context and, in one instance, by misquotation."[39] He was being scapegoated (or so he believed).

To US State Department officials the disaster represented an opportunity instead. It showed that time had come for NATO nations to invest more in environmental research. This was in line with the agenda of US President Lyndon Johnson too, as already in 1965 he had called a meeting of leading experts to advise him on how to tackle environmental degradation.[40] Furthermore, NATO officials may not have noted yet, but Rabi certainly did, that the scientists sponsored by the alliance could put together relevant knowledge. For instance, the British oceanographer and member of NATO subcommittee on oceanography George Deacon had resented being left out of Zuckerman's task force because a study of the channel's currents could have contributed to limiting the spill.[41] And the Belgian André Capart, newly appointed chair of the subcommittee, had lobbied for support on research to prevent oil pollution.[42]

On March 21, 1968, on the occasion of its tenth anniversary celebration, environmental degradation moved to the center of the Science Committee's debate. As anticipated by McLucas, the anniversary celebrations were not cheerful. Brosio opened proceedings, shortly followed by the Norwegian "Wise Man" Halvard Lange.[43] Then Zuckerman's speech went down like a lead balloon. Still embittered by the *Torrey Canyon* affair, he recalled only the committee's troubles. He was curt in saying that "the Science Committee was given little opportunity to help" in the organization of defense science. He did not criticize committee members, but he did not pay tribute to their achievements either.[44] He painted a disheartening portrait.

In a somewhat ironic twist, Rabi now put environmental issues at the center of his own recollections. The science diplomat transformed the main reason for Zuckerman's bitterness into something that could open a bright future for the committee. He did not echo Zuckerman's polemic on defense research, nor did he recall the recent outbursts with Brosio, on appointing a science adviser, or the tensions with Giacomini on science and development. He stressed only that the problems of the alliance had grown increasingly complex and that science could offer solutions: "I refer to such matters as pollution of the environment: air, water, soil, noise, congestion—both of living space and travel." An increasingly impoverished environment was the "inevitable consequences of the heedless application of science and technology," he argued.[45]

To some extent Rabi's reflection was inspired by an environmental sentiment. Rachel Carson's influential 1962 bestseller *Silent Spring* had pitted its marine-biologist-turned-author against the chemical giant DuPont. Conservationists such as Barry Commoner had by then gained popularity in the United States.[46] Environmentalism had exploded on US university campuses too, in conjunction with anti–Vietnam war protests. Days before Rabi's 1968 speech at NATO headquarters, the students of New York University had rallied against Dow Chemical, the chief US manufacturer of the notorious Agent Orange. And a few days later, Rabi's students at Columbia University contributed to the disruption of the Democratic National Convention in Chicago.[47] Environmentalism had also carved a new space for political action for otherwise marginalized groups of nature conservationists. To the UK ornithologist-turned-environmentalist Max Nicholson, for instance, the designing of new environmental schemes was tantamount to a political revolution.[48]

Rabi was also aware that environmentalism had gained appeal not just among grassroots protesters but also in the corridors of power. In his speech at the NATO anniversary, he recalled the words of British economist Barbara Ward (recently employed at Columbia University).[49] In her 1966 *Spaceship Earth*, she described human beings as both the earth's crew and its passengers. She thus underscored the responsibilities associated with steering the blue planet in the right direction.[50] Casting aside the dif-

FIGURE 4.1. Gunnar Randers (third from right) and NATO Secretary General Manlio Brosio (second from right) at NATO headquarters on the occasion of the first meeting of the Committee on the Challenges of Modern Society on December 8, 1969. Source: NATO Archives.

ficulties of the alliance, Rabi concluded his speech by stating that science "had transformed our environment almost beyond recognition and promises to continue this transformation at an ever accelerating rate."[51]

Rabi's words might have even convinced Brosio to appoint a new science adviser, for, by summer 1968, the right person for the job was found. The Norwegian Gunnar Randers was the first non-US NATO science adviser (figure 4.1), and his career trajectory was similar to Rabi's. Trained in astrophysics, he had moved to the United States (and worked with Zuckerman's deceased friend Howard P. Robertson) during World War II before returning to Europe to set up the Technical Committee of the Norwegian High Command in exile (the NDRE precursor).[52] After the war, Randers was involved in the development of the Norwegian atomic energy project, which allowed him to feature prominently in international organizations such as the International Atomic Energy Agency (IAEA).

Although accustomed to traditional NATO areas of expertise, Randers was equally interested in environmental issues because of the impacts of atomic energy programs.[53] His appointment signaled the beginning of a period of change in the Science Committee, now veering more decisively toward the promotion of environmental studies. Rabi thus succeeded in marginalizing alternative proposals for the committee's future investment and reducing the influence of those opposed to increasing the committee's budget (i.e., France and Britain). Meanwhile, US President Richard Nixon's initiative further extended the alliance's "greening" through the creation of another committee devoted to environmental actions, which marked the appearance on the NATO scene of new experts in environmental management, such as the celebrated US conservationist Russell Train.

Train and the "Action" Committee on the Challenges of Modern Society

Up to his election, the anticommunist Republican Richard Nixon had shown little interest for the environment. While campaigning for elections, however, he realized that nature's conservation had featured prominently in the political agenda of radicals, hippies, and antiwar demonstrators. And its appeal among the moderate electorate had grown too, even if just in the diluted form of desperate housewives (and their husbands) complaining about traffic congestion when driving downtown for shopping. In the period leading to his election, Nixon decided that sustainability should be part of his electoral program and set up the Task Force on Natural Resources and Environment, which was headed by Russell Train.[54]

Train's growing influence in Nixon's administration is revelatory of the ongoing changes in the US political landscape. A graduate of Columbia University, and a US tax judge, over the years the Republican Train had become a major force in the conservation movement. In 1961, he contributed to the establishment of the World Wildlife Fund and was later appointed as president of the Conservation Foundation. He also chaired the US Congress commission in charge of water management. These activities led him to engage internationally too, for instance, through contributing to the International Union for the Conservation of Nature (established in 1948), in which the Briton Max Nicholson also was active.[55]

After his appointment in the task force Train was called in to advise Nixon on environmental policies. And either out of genuine intentions or expediency, the new president successfully introduced new environmental legislation. He set up the Council for Environmental Quality (also chaired by Train) and promoted the new National Environmental Policy Act. The US president also set provisions for controlling the emission of hazardous pollutants,[56] and, at the State of the Union address of January 1,

1971, referred to environmental protection as one of the key goals of the administration.[57]

In 1969, Nixon decided to take environmentalism to NATO. Using another anniversary as a platform (that of the twentieth year of its foundation), the new president urged US allies "to explore ways in which the experience and resources of the Western nations could most effectively be marshalled toward improving the quality of life."[58] Indirectly recalling what Rabi had also argued for the previous year, Nixon thus sponsored a NATO investment in actions capable of granting such improvement. Nixon initially enlisted the Democrat and director of Harvard-MIT's Joint Center for Urban Studies Daniel Patrick Moynihan to take care of NATO environmental affairs.[59] And the newly appointed Moynihan soon headed a twenty-person US delegation to Brussels in order to impress upon allies the importance of embracing NATO's new environmental agenda (Moynihan is the US delegate speaking in figure 4.1).

The initiative initially confused delegations and, at one NAC meeting, the UK representative expressed doubts about the role that NATO should be playing, especially because it was primarily a defense alliance and other international organizations should have taken responsibility for environmental problems.[60] But with the allies criticizing the United States for the Vietnam conflict, the French rejecting military integration, the Italians on the warpath over the technology gap, and the Germans and Belgians wanting to establish an East-West dialogue (or making of NATO a force for peace, even), Nixon reckoned that shifting the focus of ongoing talks to environmental issues would be the best possible way to evade thorny questions tarnishing NATO relations.

A few months later, the CCMS was born and ready to start proceedings. On November 6, 1969, the NAC agreed to establish the committee, clarified its terms of reference, and ratified the NATO science adviser Gunnar Randers as its chair.[61] Within a month, the new committee met for the first time. Aware of the reluctance of some NATO countries (and especially Britain and France) to raise the alliance's budget, US officials now proposed that the CCMS adopt a mechanism of financial commitment different from that utilized in the NATO science program. Participation would be entirely voluntary, and delegations of individual countries could propose pilot projects on specific topics and pay for their completion (other nations could, if they wanted, cooperate as "co-pilots"). In promoting the environmental program, the NATO Scientific Affairs Division underscored its practical nature, short-term impact, and freedom from additional costs. The division bore the burden of routine secretarial work.[62]

The role Randers played as chairman of both the Science Committee and the CCMS energized the activities of both. The new mechanism of voluntary financial assistance actually convinced him that the funds available to the NATO science program should have made it far more effective. As Hamblin has argued, Randers was not an ecologist, but his words chimed with Rabi's rhetoric about the need for man to reconcile with nature.[63]

At the end of 1970, Moynihan resigned from his role as chief delegate, unhappy about the broadening range of topics that the US administration wanted the CCMS to tackle, and embittered by polemics at home.[64] He was now replaced by Russell Train, who, with his experience in environmental affairs internationally, was at greater ease than his predecessor in dealing with the officials of other NATO delegations.

By the time Train was appointed, eight CCMS pilot projects had begun, including those dealing with air and sea pollution, pollution of rivers and lakes, assistance in disaster such as earthquakes, and traffic congestion. One of the first initiatives taken by Train was to propose that the Ford Foundation sponsor a CCMS environmental fellowship program.[65] He did not succeed in securing the funding, but the US government eventually granted sponsorship for the scheme. NATO thus awarded forty-six CCMS fellowships in the first five years of the alliance's environmental program, and the project covered subjects as diverse as the role of experts in environmental problems, the harmonization of national environmental regulations, and mathematical modeling of growth and water waste.[66]

Some of the pilot projects extended international collaboration on specific environmental issues. For instance, that on air pollution brought together environmental analysis teams from the United States, Turkey, and West Germany to compare air quality in the urban conglomerates of Ankara, Saint Louis, and Frankfurt. Their final report emphasized the importance of a more accurate understanding of meteorological processes in urban spaces and the need for better monitoring and modeling techniques. Notably, the completion of these projects revealed for the first time the potential for remote-sensing techniques elaborated in the context of defense-oriented NATO studies to be applied to environmental monitoring tasks. In November 1971 an air-quality resolution was submitted to the NAC for approval.[67]

Because the *Torrey Canyon* incident was important in instigating NATO's transition to environmentalism, the CCMS sea pollution pilot project focused especially on oil spills. The initiative was also responding to another environmental disaster. In January 1969, a blowout in Platform A off the Santa Barbara channel (in southern California) caused another major oil leakage—the largest recorded in US waters to that date.[68] Piloted by the

Belgian delegation, the project led to the organization of a meeting of experts held in Brussels on November 2–6, 1970. One hundred representatives of fourteen NATO countries and observers from Japan, Australia, and Spain attended it. The participants called for an initiative to abate oil spills and called for their elimination by 1975.[69] The NAC officials now agreed to propose a new convention replacing the existing provisions dating back to 1954 (Convention on the Prevention of Pollution at Sea by Oil) and also stipulated that NATO inform the activities of other international organizations responsible for these matters, such as the Intergovernmental Maritime Consultative Organization (IMCO) and the WMO Joint Group of Experts on Scientific Aspects of Marine Pollution.[70]

To enhance Nixon's initiative, Train, assisted by US State Department official Harry C. Blaney, instructed that every CCMS-related activity should be promptly advertised so that the media could report on it. Before each CCMS meeting, the US Information Agency (USIA) coordinated with the NATO Information Service (NATIS) to prepare press kits, tapes, film sessions, and much more for press distribution.[71] Moreover, before and after CCMS meetings, the USIA (and Train personally) took responsibility for preparing letters to send to the directors of major newspapers to secure their support. Media coverage was carefully monitored too, including the articles of the *Washington Post* correspondent Alfred Friendly (whom Train nicknamed "the unfriendly Friendly") that had negatively reported on the CCMS and revealed the lack of agreement between allies on NATO's third dimension.[72] Moreover, the CCMS meetings often left journalists puzzled and, on one occasion, one US diplomat anxiously reported that "most newsmen came away scratching their heads, either not feeling there was a real story in the meeting, or not being able to put their finger on what their lead should be."[73]

In 1974, Train even agreed to be part of a media stunt in order to impress the befuddled commentators and arranged to present a recently launched US pilot project on clean automotive systems by driving a small electric-powered vehicle in the parking lot just outside NATO's headquarters together with the new secretary general, Joseph Luns. The journalistic stunt worked only to some extent in that the prototype six-foot-six vehicle resembled a golf cart was difficult to use and drive. The NATO photographic film produced on that day shows the environmentalist and the secretary general rather apprehensive about getting into the narrow vehicle. They needed assistance in order to get into the car because the hood had to be lifted to get in. They were wary about its narrow seating, and they needed instructions about how to drive it. They eventually showed journalists

FIGURE 4.2. An apprehensive Luns (right, with Train, left) receives instruction on how to drive the electric car. Source: NATO Archives.

the merit by driving the super-compact car as scheduled, but at the end of the promotional exercise they struggled again to lift the hood and get out of the car (see figures 4.2, 4.3, and 4.4).

Two years after CCMS's establishment, Randers recalled the committee's achievements with enthusiasm.[74] Meanwhile, the US delegation at NATO continued to reassure State Department officials that the original skepticism about the CCMS had largely dissipated. And, to some extent, the committee bore the diplomatic fruits that US officials had hoped for, especially in that the French enthusiastically endorsed the CCMS concept although out of NATO's military command. West German delegates also wholeheartedly agreed to support Nixon's initiative, thus strengthening the alliance as a whole and persuading other delegations.[75] The CCMS launch also gave US diplomats an opportunity to foster relations with other administrations. Because the Italians had been anxious about the growing technology gap, from 1971, the US delegation at NATO continued to push for support of an Italian pilot project on disaster assistance.[76] The CCMS also paved the way to more collaborative plans even across Cold War divides. On May 23, 1972, the US and the USSR administrations signed a bilateral agreement on Cooperation on Environmental Protection.

But the optimism exuding from US cables assembled in embassies,

FIGURE 4.3   Off they go! Luns and Train in the NATO headquarters parking lot. Source: NATO Archives.

FIGURE 4.4.   Luns struggles to lift the hood and get out of the car. Source: NATO.

consulates, the State Department office, the White House, and other centers of US diplomacy waned fairly soon. As the next chapter will show, the impressive effort to persuade allies and non-NATO nations to jump on the environmental bandwagon was less successful than expected. As Luns and Train were busy promoting the electric-engine car at NATO, Nixon's adviser Henry Kissinger instructed Train to put the CCMS under review. But evaluating the committee only on the basis of its achievements would be misleading. It was not born to revolutionize the alliance, but to stabilize it, actually. And it was mainly the Science Committee that benefited from this rejuvenation effort.

Pestel and the Science Committee's Environmental Turn

On October 6, 1969, the Science Committee met for the first time in Rabi's backyard. Welcomed at the US State Department by the Undersecretary of State Elliott L. Richardson, its members also were greeted by Nixon's advisers on environmental affairs, including Daniel Patrick Moynihan and Russell Train.

Rabi, Moynihan, and Train now found an important ally in their efforts to persuade NATO delegates that the environment should take center stage in the committee's activities as the West German representative Eduard C. Pestel openly endorsed their plans for sponsoring environmental research through NATO. Pestel's appointment as German delegate actually highlighted the wish of West Germany's Christian Democrat administration to play a more prominent role in NATO's scientific affairs, especially now that West Germany contributed a larger share of the program's budget. Pestel retained his role as West German representative even when the socialist former mayor of West Berlin Willy Brandt was appointed chancellor. On top of being a successful researcher, Pestel had previous experience as a science administrator. An industrial designer by profession and professor of mechanics at the Technical University of Hannover, Pestel had previously played a leading role in the federal research foundation *Deutsche Forschungsgemeinschaft*.[77]

Pestel was sympathetic with Rabi's pledge also because of his recent studies on environmental degradation. In 1968, he had contributed to the proceedings of the Club of Rome, one of the most prominent scholarly associations pioneering sustainability studies.[78] The community of intents between Pestel and Rabi also echoed the search for broader political synergies between the US and West German administrations. The Germans were about to complete paying back the loan received under the European recovery program, and their standing in NATO was further heightened by being permanent members of the NATO Nuclear Planning Group.

During the Science Committee meeting of October 1969, Rabi, Randers, and Pestel called for a substantial increase in the committee's budget so as to push environmental research firmly in its agenda. Moynihan now stressed that the committee would be well placed to sponsor environmental studies, and Pestel explained how environmental planning could harmonize technological development and social requirements. NATO-sponsored scientists could advise on this harmonization, Rabi concluded.[79] The committee eventually agreed to open a separate budget line for program development so as to upgrade the research grants program in line with NATO's greening.[80]

But Pestel's and Rabi's crafty diplomatic work found insufficient support at this stage. Their words were met with silence, especially by the representatives of the two countries, Italy and the United Kingdom, that now paid the largest share of NATO's science budget. Their representatives' dissent over an environmental focus actually paralleled a plan to invest more in alternative setups. By 1969, these officials had embraced Fanfani's early pledge of support to science and technology by suggesting that the two countries pool resources, especially in nuclear and missile engineering, by establishing an Anglo-Italian Committee on Science and Technology. The committee, however, never really took off. It met the first time on January 14, 1970, but Wilson exited office the following June, and his Italian counterpart, Mariano Rumor, resigned shortly after. Their resignation coincided in both countries with a comprehensive review of policy, and in Britain with the first significant opening toward sponsoring environmental studies.[81] Still unhappy about NATO science, the British now turned (once again) to the French for support. The conservative Prime Minister Edward Heath worked toward improving relations with France, which in turn lifted the veto on Britain joining the European market and portended collaboration on nuclear energy. Even this prospect never materialized, however, as Nixon undermined a nuclear *entente cordiale* by offering ad hoc cooperation to both countries (delivery systems to the French and a replacement for Polaris to the Britons).[82]

Notwithstanding the divisive forces operating within, the Science Committee fully endorsed environmentalism. The media resonance that NATO's environmental program received eventually brought grist to Rabi's and Pestel's mill. The proposition that the NATO Scientific Affairs Division be used as the CCMS secretariat signaled that the division could be dealing with more than the committee's routine business and that the

CCMS model of sponsorship could be implemented with no additional costs. Furthermore, US officials requested that the members contribute to committee activities when suitable, for instance, by advising on how to tackle sea pollution problems and sponsoring research on "a tracking system for oil tankers."[83] At the end of 1970, the US delegation made an explicit request that "methods of continuing surveillance of oil spills" be developed.[84]

These propositions remind us that the transition in patronage here described facilitated an important shift in environmental research practices. After the *Torrey Canyon* incident, Rabi, Pestel, and other committee members foresaw that expertise developed while tackling traditional Cold War surveillance tasks could now be redeployed. In particular, they anticipated the merits to be derived from using remote-sensing approaches in remedial environmental actions. In so doing, they envisaged sponsorship of environmental research that would have invested directly the scientists who were recipients of NATO grants.[85] Redeployment also produced synergies between experts who had been involved in nature conservancy and adopted ecological approaches and those who embedded the environmental discourse in surveillance-oriented studies.

The committee's quarrel on budget continued for one more year, at the end of which its members finally agreed on the sums to be disbursed for the 1971 program, taking the ceiling for science fellowships to 2.7 million US dollars, those for the ASIs to 890,000 and that for research grants to 570,000. Because the subcommittee on oceanographic research was resistant to calls for cutting costs, the committee members agreed to separate its budget from that on research grants so as to keep their expenditure within a maximum ceiling.[86]

Randers sought a compromise between those, such as German and US representatives, who advocated a radical change in NATO's sponsorship (and wanted to invest more) and those (British, French, and LDC delegates) who resisted change (albeit for different reasons). This led him even to confront colleagues in Norway, including the former chair of the subcommittee on oceanography Håkon Mosby, who had plans for costly oceanography projects. But the search for an environmental agenda continued and now found important allies outside NATO.

In February 1971, Carroll Wilson, professor of management at the MIT Sloan School and director of the Study of Critical Environmental Problems (SCEP) group, was invited to attend the next Science Committee meeting. Wilson's expertise in management bespeaks his education in operational methods, but it was especially his recent work that attracted the committee's members. After the experience with the Club of Rome (in which

Pestel had also been active), Wilson assembled a group of MIT experts with funding from the Ford, Rockefeller, and Sloan foundations. The result was *Man's Impact on the Global Environment Assessment and Recommendations for Action*, a study laying out for the first time plans for a global environmental monitoring scheme.[87] Remote-sensing-based methods were now propagandized as the way forward in addressing environmental degradation and the Global Environmental Monitoring System was eventually adopted in the context of the UN Environmental Program.[88]

Wilson emphasized the need to chart atmospheric change and, for the first time, recalled an item that had already captured the media attention: anthropogenic climate change. Presenting new data on the rising concentration of carbon dioxide in the atmosphere and the dangers associated with its effects on climate, Wilson posited that carbon dioxide should now be more closely monitored. Pestel now proposed that the committee directly contribute to the effort. But responses were mixed. Lied warned that Pestel's proposal amounted to "succumbing to fashionable trends."[89] Rabi's alternate and former NATO science adviser, William Nierenberg, was concerned about the alliance's commitment to other large projects in oceanography. In 1971, the Science Committee gave mandate to Pestel to set up a small working group also comprising Cottrell (see figure 4.5), Alves Martins, and Rabi. The group now laid out a new structure for the Science Committee and its subsidiary bodies. Pestel's working group emphasized the need to set up environmentally focused panels and advocated that one or more members of the Science Committee be involved in these panels in order to contribute more to its activities.[90]

The Science Committee stipulated that the overall allocation for NATO science (5.8 million US dollars) be divided among fellowships (50%) and ASIs (20%) and that the remaining 30 percent be divided among the funding of individual projects (15%), old groups from previous years (human factors, oceanography, radio meteorology), and new ones (eco-sciences).[91] From 1973, two other new panels devoted to air-sea interaction and marine sciences began operating with budgets just below one hundred thousand dollars each. It is important to recall that the funds available for these new environmentally oriented panels still corresponded to between one-third and one-half of the overall budget for the NATO research grants program. This distribution confirmed the focus on environmental studies distinctive to NATO's science program, although now geared toward the study of environmental degradation (and, overall, a more conservative budget).

The establishment of the eco-sciences panel marked a substantial change in the management of the Science Committee's subgroups. On September 8, 1971, eight experts were invited to NATO headquarters to

FIGURE 4.5. An unusual panoramic shot of a Science Committee meeting in May 1973. Seated: Alan Cottrell (UK), Gunnar Randers (chairman), L. Néel (France). Standing: T. Karlsson (Iceland), C. Guerra (Portugal), N. Özdas (Turkey), A. A. Th. M. van Trier (Netherlands), J. R. Whitehead (Canada), W. Nierenberg (USA), E. Pestel (West Germany), Finn Lied (Norway), P. G. Jensen (Denmark), P. S. Theocaris (Greece), A. P. Boever (Luxembourg, obscured), and F. Cerulus (Belgium). The Italian delegate Amedeo Giacomini was absent. Source: NATO Archives.

confer about setting it up. Randers arranged the meeting in collaboration with his deputy, the former US State Department official Eugene Kovach. In introducing the new panel, Randers recalled that the group was a "somewhat different" one in that it connected to CCMS activities.[92]

The Scottish-born Canadian Patrick D. McTaggart-Cowan, president of Simon Fraser University (British Columbia, Canada) and formerly member of the Meteorological Service of Canada, was appointed the eco-science panel's first chairman. His career epitomized the changes now occurring at NATO. An expert on the polar jet streams, in 1967 McTaggart-Cowan had become more concerned with colonization of the far north and also advocated the search for ways to harmonize colonization and environmental change.[93] His panel focused especially on the monitoring of pollutants in the atmosphere and the sea, and a subpanel was established to look into toxic sea substances. It was chaired by the British fishery scientist Alasdair McIntyre of the Torry Marine Research Station in Aberdeen, Scotland.[94] The US marine scientist Carl Henry Oppenheimer, head of the Department of Oceanography at Florida State University, also was invited to contribute to the new panel, and he promoted research on the ecological impact of solid waste, in line with his studies on bioremediation and the ecology of oil-polluted areas.[95]

Toxicology, recovery of devastated ecosystems and environmental monitoring became the three cornerstones of the eco-sciences panel.[96] From 1974, the panel also agreed to promote a study of acid rains and appointed the Canadian climatologist Frederick Kenneth Hare, who was another of the speakers at the AGARD 1956 *Polar Atmosphere* symposium.[97] As one might expect, the shift in emphasis in NATO's program chimed with the career change of some of its contributors. On the occasion of the Science Committee meeting of May 21, 1973, Randers reflected on the changes that had occurred since 1968, when he was appointed science adviser, and remarked that the science program had now become more independent from military applications.[98]

This chapter has explained this transition in light of US diplomacy efforts to evade key issues marring the cohesiveness of the alliance, especially those on the nature and functions of the alliance and its role in the next phase of the Cold War. It also has shed new light on the search for new research themes that would restore consensus within the Science Committee after the disagreement between two of the administrations that had sponsored its creation (US and UK) had grown considerably.

NATO's greening thus resulted from the shortcomings of the alliance's initial science diplomacy efforts, which was in itself a way to address those of its traditional diplomacy initiatives. This chapter has shown these deficiencies to include the lack of endorsement of NATO's science program by its military authorities and the oppositions to budget increases by Britain and France. It has also introduced an additional controversy in connection with requests by LDCs' representatives to tie the science program to a development agenda. This further eroded NATO relations and contributed to the shift from science to environmental diplomacy. The diplomacy effort bore fruits and certainly benefited from the active participation in its design by leading science diplomats like Rabi, who appreciated the importance of the science program for the alliance's affairs, and nature conservationists like Train, who had experience in the promotion of environmental programs internationally.

The US allies' commitment to the new environmental research and action programs created the circumstances for temporarily rebuilding consensus at NATO and, crucially for the argument of this book, NATO's "greening" led to the introduction of traditional techniques of remote sensing into environmental studies. We shall see in chapter 6 that in the 1970s and 1980s NATO continued to inform both defense- and environmental-protection-oriented studies with remote sensing to the point that their promoters found opportunities to collaborate and share ideas.

The next chapter focuses instead on the limits of NATO's entanglement of science, environmentalism, and diplomacy. Because diplomacy was the hidden ambition of NATO's environmental actions, when these actions affected relations between allies more, they were promptly discarded; something that was decisive in defining the trajectory of both the CCMS and the Science Committee in the 1970s. All the while inflation started gnawing at NATO's investment.

FIVE

NATO's "Greening" in the Decade of Inflation

The degradation of the environment is an international problem which requires multi-national efforts [. . .].[1]

Yet today NATO does not even require its contractors to put out environmental impact statements.[2]

If the tenth anniversary of the Science Committee had been an opportunity to begrudgingly recall its past activities, then the twentieth anniversary meant to cast a positive light on the committee's future. On April 11–13, 1978, its members were invited to meet in the luxurious Palais d'Egmont (see figures 5.1 and 5.2); a neoclassical mansion in central Brussels miles away from the barbed-wired NATO headquarters. This time the criticism-prone Zuckerman was in attendance, but he was not on the list of speakers. The NATO Secretary General Joseph Luns agreed that Zuckerman's old acquaintance Rabi should briefly offer some reflections on the committee's past together with the Nobel Laureate Néel.

After their speeches, the discussion moved on to consider future challenges to the Western scientific community. Among the speakers were the Club of Rome's figurehead, Augusto Peccei, and the MIT political scientist Eugene Skolnikoff. The honor of shedding light on NATO's new science and environmental programs was left to Eduard Pestel.[3] The choice of Pestel as last speaker meant to reiterate the importance of the recent changes occurring in NATO's science

FIGURE 5.1. Still friends after all these years? Zuckerman (center) and Rabi (right) during the celebrations at the *Palais d'Egmont*. Source: NATO Archives.

program. As the last chapter showed, the German representative, together with Rabi, took responsibility for reorienting NATO science so as to prioritize environmental research.

But NATO's greening—if intended as an alignment to nature conservation research and actions—yielded less than expected, though, especially when inflation bit the economies of its member states harder. Oil-driven price rises made it more difficult to set up new research projects with the sums that the alliance could disburse for novel research. These circumstances did not prevent Rabi and Pestel from reforming NATO's science program, although, as more parsimonious environmental research groups replaced the profligate ones born in post-Sputnik times. The Science Committee members now agreed that a new generation of scientists prone to investigate environmental degradation more firmly move into the alliance's sponsorship initiatives. Veteran recipients either agreed to shift research focus or had to find new sponsors. Contingently, new subgroups stimulated synergies with the CCMS action panels, while those on oceanography and meteorology were disbanded. NATO's new environmental focus was decisive in defining a new consensus among committee members.

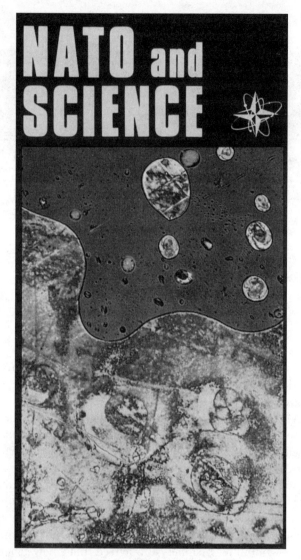

FIGURE 5.2. The quirky cover of the NATO leaflet distributed on the occasion of the celebrations. Source: NATO Archives.

Environmental diplomacy did not yield enough in the context of CCMS activities, and not just because of inflation. National delegates continued to have reservations about NATO's environmental focus because it deeply affected the alliance's relations, especially when new actions against pollution went against the interests of powerful industrial groups.

Establishing the CCMS had been an exercise in political expediency aiming to quell competing plans and evade divisive issues, but keeping the committee running proved harder. Because delegations were often reluctant to commit, decisions on key pieces of international legislation, such as a new convention on sea pollution, were seriously delayed. Moreover, NATO continued to overlook critical environmental issues directly relevant to its roles and functions, such as the impacts of its military wastes and exercises. By 1974, with Nixon just about to leave office, it was even unclear whether the CCMS should continue to exist.

In these difficult circumstances, the collaboration between the Science Committee and the CCMS established a distinctive approach to environmental management and policy. NATO actions conformed to what John Dryzek has defined as "prometheanism," an understanding of environmental problems as shortcomings in the process of adapting nature to the needs of mankind.[4] Worthy of notice, for instance, is the talk of Canadian climatologist Frederick Kenneth Hare at the twentieth anniversary of the Science Committee's foundation. Typified by an unjustified and Western-centric optimistic note, Hare's speech recalled that NATO countries had already made a major leap forward in adopting environmental practices. Other countries should follow suit to achieve similar results.[5] Techno-optimism also informed the reuse of traditional remote-sensing methods, previously elaborated in the context of traditional defense research exercises, in order to chart changes in degraded natural environments.[6]

And so, during the 1970s, NATO's science and environmental programs continued to be a source of dramatic contradictions. On the one hand, the alliance continued to be a major contributor to contemporary debates on environmental protection. But, on the other, it cast its role as very different—alternative even—to that of those nongovernmental organizations devoted to nature conservation that by then had become more prominent in NATO countries.

Challenges to the Challenges of Modern Society Committee

In the first half of the 1970s, US envoys at NATO continued to display confidence about CCMS activities and trajectory, especially because they paralleled outstanding developments on the international scene. The first Earth Day took place on April 22, 1970. Proposed by US senator Gaylord Nelson (Wisconsin), it aimed to raise awareness of environmental themes within the civil society and education institutions.[7] At the level of international policymaking, developments were equally important; UN

General Assembly Resolution 2398 called for the organization of a conference on the human environment. Stockholm was chosen as the site of the worldwide environmental meeting.

But views on environmental policy differed quite dramatically at NATO headquarters. Back in the early 1970s, the former NATO secretary general Brosio had often complained about CCMS procedures and the organization's other serious shortcomings: "sluggish progress, imprecise and generalized recommendations, emphasis on research as opposed to action, lack of clearly defined goals, overlap with effort of other organizations."[8] Brosio praised the CCMS only for the introduction of the pilot-project mechanism, which he judged innovative. He was otherwise doubtful about the committee's role in NATO affairs. Environmental propositions destabilized the alliance, as he confided to a Dutch diplomat, especially since attempts to strengthen the CCMS alienated delegates that opposed the committee.[9]

The CCMS initiative had actually found NATO allies very passive. By 1972, the United States had piloted three projects and was co-piloting one more. But the responses from Italy and the United Kingdom were weaker, for their national groups agreed to pilot only one project each. The initiative had generated enthusiasm in West Germany, France, and Canada. These countries were now piloting or co-piloting no fewer than three projects. The Norwegians, however, had been less eager to contribute, and their negative stance was shared by the Danish and Dutch delegations. Their reluctance originated not from lack of environmental sentiment, but from the belief that NATO was not the right forum to tackle pollution. US officials deliberately overlooked the Norwegian skepticism when reporting to Washington, DC, and satisfactorily noted the involvement of Turkey, Portugal, and Greece instead.[10]

In 1971, Train prepared a cheerful summary for Nixon, suggesting that there was sufficient support for the CCMS at NATO headquarters.[11] But even his most trusted collaborator at the US State Department had doubts. Harry C. Blaney had just filed a report for Train's eyes only, which anticipated his resignation. In the report he stressed "the inability of some of our Allies to provide creative and strong leadership" to CCMS initiatives.[12] The following year, he vented his criticism on the pages of the *Atlantic Community* and reiterated "that many NATO Ambassadors and their Foreign Offices have little appreciation or understanding of what the CCMS is all about."[13]

Did NATO diplomats know about the implications of CCMS work? They certainly worried about CCMS initiatives that ran counter those of powerful economic interests in their own countries. The French and Britons in particular were anxious about the proposal for an international convention on sea pollution that, after the *SS Torrey Canyon* and Santa Barbara disasters, was supposed to proactively tackle oil spills. A piece of legislation on the proposal had originally been proposed by the members of the CCMS pilot project on sea pollution (which had elaborated it after a NATO symposium on oil spills of November 1970) and eventually put before the NAC for approval.

While the original proposal was being examined by the council, the US delegation put together the relevant evidence showing that to seriously deal with oil spillage meant much more than just preventing *Torrey Canyon*-like disasters. One month before the NAC meeting of November 1970, a US official wrote to Train that a significant portion of the oil leaked in the sea resulted from the practice of deballasting, namely the process by which a ship increases and decreases buoyancy through taking in and letting out seawater. Because separate water tanks for cargo and ballast did not exist in ships built before 1970, up to three million tons of oil annually poured into the world oceans in routine deballasting operations.[14]

No convention on abating oil spills was bound to be successful unless deballasting was addressed. NATO's convention set out in December 1970 endorsed in its technical details the principle that oil-carrying ships should have separate cargo and ballast tanks. But when Train and some of his peers in other countries put these measures before the representatives of the shipping industry, the diplomats were openly challenged.[15] In April 1971, Train and Blaney met with the president of the American Institute of Merchant Shipping, who confirmed that shipping entrepreneurs viewed the resolution as "beyond attainment."[16]

By then, the NAC had approved a new convention on sea pollution, although in the NATO press release the question of implementations was deliberately left in the background so as to avoid upsetting oil-shipping concerns.[17] Just before resigning, Blaney urged Train to lobby with "certain interested governments" for the swift implementation of the amended oil spills resolution and stressed that Britain was "key in this context" because its representatives were against the concept of segregated ballast tanks.[18] The position of the UK delegation had emerged after consultations between the UK Foreign and Commonwealth Office and the Department of Trade and Industry, which, in light of the shipping interests represented within, had solicited resistance to far-reaching legislation. The UK delegation had already advocated the adoption of an equally viable, but less radical, environmental management provision, *load-on-top*, also sponsored by the French. It consisted of transferring the ballast mixture to a slop tank so that water and oil could separate before release. Train judged it insufficient.[19]

The international authority regulating sea navigation matters, the Maritime Safety Committee of IMCO, eventually examined the NATO convention. And since IMCO had far less weight in enforcing new provisions, those NATO delegations opposed to radical solutions to tackle oil spills succeeded in having them revised to accommodate the wishes of their shipping industries. Although load-on-top was not adopted, deballasting principles were enforced only for new ships of deadweight above seventy thousand tons; old and smaller vessels would thus be left free to pollute.[20]

Furthermore, although on November 2, 1973, IMCO agreed to the NATO convention on oil spills, thus launching the International Convention for the Prevention of Pollution, the new provisions stalled in its offices for several years. To be implemented, the IMCO convention needed to be signed and ratified by fifteen countries with a combined merchant fleet amounting to 50 percent of world shipping. Several allies did not sign it, and by 1976 only three non-NATO countries had ratified it (Jordan, Kenya, and Tunisia). So two years later the yet-to-be-approved convention originating at NATO was absorbed in the parent MARPOL protocol. It now set far more stringent measures, but notwithstanding the 1970 NATO resolution calling for a ban "by 1975 if possible, but not later than the end of the decade," MARPOL entered into force only in 1983, allowing three more years for parties to the Convention to implement the regulations.[21]

Of course these delays were the result of issues that NATO officials could not, presumably, directly address. But the implementation of recommendations proposed at CCMS meetings proved equally problematic. In 1973, Blaney's successor at the CEQ, Frank Hodsoll, lamented that only the United States, West Germany, and Canada had provided follow-up reports on the compliance of their shipping industries with the NATO protocol, notwithstanding the fact that the head of the sea pollution pilot project had already requested these reports four times.[22] Moreover, US efforts to extend the oil spills convention to ocean dumping met the opposition of British, Canadian, and French negotiators.[23]

Other CCMS proposals never saw the light of day. Hodsoll stressed that when the Canadians presented a project on stratospheric air pollution caused by air traffic, the Foreign and Commonwealth Office instructed its British representatives to "vehemently" oppose the project in light of the launch of the new supersonic carrier jet Concorde. They even mobilized the UK Meteorological Office to demonstrate that increased emissions of nitrate and sulfate oxides would cause no harm to the stratosphere's chemical composition.[24] The International Road Safety Resolution proposed by the Germans made British officials "waiver."[25] Other countries' positioning was equally difficult to predict. Norwegians and Danes agreed to unilaterally stop ocean dumping, but US officials feared that they "may be limiting their proposal to the North Sea." Developed countries agreed on a dumping ban, but only because, allegedly, they already dumped "their wastes in the LDCs' waters" instead.[26] When Train and Blaney designed the media stunt aiming to persuade allies to invest more in clean-engine cars, one of their colleagues at the State Department registered the anxiety of Western European delegates. They worried that the initiative actually aimed to drive the European automotive industry out of the US market.[27]

In his reports to Nixon, Train attempted to hide the fact that NATO allies dragged their feet and sought to emphasize his efforts to increase participation. For instance, although in August 1972 a UK pilot project on work satisfaction was about to be terminated because it had received criticism by the British trade unions,[28] Train urged Cottrell to make sure that the decision to terminate the project would not "prejudice your country's participation in other CCMS endeavors."[29]

On the whole, CCMS initiatives appeared to some delegates to be too elitist, "establishment-like." They held that NATO governments should act for the well-being of their citizens rather than for environmental sustainability alone. And so, the emphasis on traffic congestion and safer vehicles was tailored to address the problems of advanced NATO nations in particular rather than concerns about the state of the environment. The CCMS even rejected taking part in the UN-sponsored 1972 Stockholm conference that led to the establishment of a UN environmental program. At one of its meetings, the convened CCMS delegates concluded that attendance would "remove attention from other important international activities."[30] Randers also contended that the committee should not be involved in the initiative, presumably because he was anxious about the influence that Eastern Bloc countries and (nonaligned) developing nations would play in environmental affairs.

NATO's ambivalence toward tighter environmental policies severely weakened the CCMS. US representatives like Train were quick to point their fingers at their colleagues in Western Europe and their ties to local industrial concerns. In an article published in 1973, Train recalled the committee's achievements, although at the same time underscoring its shortcomings. Non-NATO governments, such as those of Japan and Sweden, were interested in CCMS projects but had to "overcome general political concern" in order to participate, and US allies continued to believe that

FIGURE 5.3. Logo on the cover of NATO leaflets advertising the CCMS activities in the 1970s. Source: NATO Archives.

NATO was the wrong forum for environmental actions and did not support such projects enough.[31]

The environmental aspects of defense-related activities represented a particularly touchy subject for US and NATO officials alike, as shown by their lack of reactions to an article published in 1973 on Canada's chief global policy review *International Journal*. The Canadian scholar J. Patrick Kyba, who was awarded a CCMS fellowship, put his finger on the most startling contradiction of NATO's environmental work. Although supportive of CCMS activities in general, Kyba argued that the alliance routinely produced pollutants in the form of ship- and shore-generated military wastes. During its exercises and through hundreds of military installations, NATO's forces had a massive impact on the environment. Why had NATO never attempted to tackle the very pollution it directly produced?[32]

Kyba's compelling question was never answered. And by 1974, NATO's environmental bubble was about to burst. By then, the financial crisis had already affected the running of the four-year-old US Environmental Protection Agency (EPA) and deteriorating economic circumstances prevented investing further in environmental actions.[33] Moreover, the growing uncertainty on the future of Nixon's administration also played a role, in that the president's illegal use of power in the Watergate scandal led to an impeachment procedure first and then forced him to resign.[34] The shake-up that followed Nixon's departure rippled through a number of US departments and hit the office responsible for the CCMS too. The US State Department now agreed to set up a new Bureau of Oceans and International Environmental and Scientific Affairs, which would take responsibility for CCMS matters too, thus pooling scientific and environmental diplomacy units in an effort to streamline and cut costs. Hodsoll believed the decision to be mistaken. US leadership in CCMS would be compromised, he claimed, "unless the CCMS officer is given a good bit of latitude and is not hamstrung by bureaucratic bickering."[35]

The US administration nearly came to fully disengage from NATO's environmental diplomacy effort. At the end of 1973, Kissinger had already asked Train to evaluate four options for the CCMS's future: boost it through renewed assurance of support from the president; promote its transfer to the OECD; call for an ad hoc meeting of environmental ministers from NATO countries; let it wither away.[36] US policy toward the CCMS was now under review, although decisions would be postponed until after celebrations for NATO's twenty-fifth anniversary.[37] On that occasion, Kissinger famously stated that the alliance "builds on the past without becoming its prisoner," which in the case of the CCMS meant that Nixon's adviser did not intend to disband the committee, but that it would not be lifted out of NATO's swamp waters either. Kissinger only wished to keep it floating in the alliance's bureaucratic mud with the minimal amount of buoyancy that financial circumstances allowed.[38]

This is exactly what happened as the CCMS continued to seek a precarious balance between the search for solutions to environmental problems and NATO's political and defense imperatives. Gerald Ford did little to generate new enthusiasm for environmental policies when he replaced Nixon as president, and Train, who in 1973 had taken control of the EPA, eventually agreed to play a somewhat less key role at NATO alternating with other US officials.[39] In his annual visits to NATO, Train continued to push for greater commitment to environmental protection; often without success. In 1976, for instance, he recalled that the 1973 Endangered Species Convention had been signed only by five out of fifteen NATO countries and urged to do more.[40]

But Nixon and Ford never intended to promote the CCMS in order to transform the alliance as a champion of environmentalism, but rather to take forward a set of actions that would shift the focus of NATO's ongoing political debate and boost the alliance's integration. In particular, the CCMS establishment fostered the reformation of the Science Committee. Rabi and Pestel could now execute their plans to transform the committee's structure to align its agenda with that of the CCMS. And Randers could argue that, because the CCMS projects were a paragon of financial efficiency, some of the science projects lavishly funded by the alliance should be abandoned. The NATO subcommittee on oceanographic research was about to be the first casualty of this reorganization.

The Ashes of Oceanography and the Phoenix of Marine Science

NATO had always prioritized the funding of oceanography, but in the 1970s the position of the subcommittee on oceanographic research had become unstable mainly as a result of its costly research program. And its budget had made it more difficult for the Science Committee to reform NATO's research grants program. In the past, its representatives had requested that the funding of the subcommittee on oceanographic research be reexamined and questioned about whether some of the projects should continue indefinitely, as those proposed by Norwegian oceanographers had received yearly renewals.[41] The members of the advisory grants panel also expressed "some concern about the large proportion of the budget which was already devoted to special fields, which greatly limited the ability of the panel to support new projects."[42] Nierenberg made patently clear that NATO oceanographers were granted untenable financial autonomy because "it is [. . .] this subcommittee [. . .] which sets the budget for the oceanographic effort."[43] What was not put on record, presumably, was that NATO oceanographers could fill grant applications with one hand and recommend their funding (as panel members) with the other. In various circumstances, the advisory grant panel openly stated that oceanography represented an "inertial mass" taking one-third of the resources available for research.[44] Using the expedient of the environmental revolution fashionable at the time at NATO, Randers solicited a revision of the subcommittee's position.

NATO-sponsored oceanographers actually wanted to embrace a new environmental agenda, especially upon the appointment of the Belgian André Capart as the new chairman. Director of the Royal Belgian Institute (and Museum) of Natural Sciences (Brussels),[45] Capart had gained a position of responsibility on the NATO subcommittee when he succeeded in readapting Norway-designed instrumentation for the measurement of currents for NATO's surveys in the Mediterranean. He also laid out plans for a floating laboratory that could promote a wider range of investigations than those sponsored by the alliance at the time, including sea pollution. Capart's research agenda had environmental aspirations, but the costs associated with the floating oceanographic laboratory he hoped to set up resembled the committee's past and present profligacy: two million US dollars to establish it and a quarter of a million per year to operate it.[46]

It is unsurprising that Capart's plans produced apprehension. In 1971, the Division of Scientific Affairs instructed the subcommittee "to work within a maximum ceiling to its total budget" after the separation of financial allocations for oceanography and the other strands of the research grants program.[47] Moreover, in the aftermath of late-1960s campus protests, oceanographic research came under greater public scrutiny. Physical oceanographers seemed too cozy with military organizations, especially because the incident of vessel *USS Pueblo* attracted media attention. The oceanographic vessel was captured by the North Korean navy with the intent of gathering signal intelligence (that is to say, spying on North Korean communications). In recalling the episode, the opening pages of the prestigious American Institute of Biological Sciences journal *Bioscience* contended that oceanography was now "prostituted by the military."[48]

NATO's headquarters was not a university campus and few, presumably, took notice. But Rabi and Randers worried about the oceanographers' failure to generate positive media attention for their endeavors. In 1970, Randers tasked Mosby, Henri Lacombe, Odd Dahl, and the US oceanographer Allyn C. Vine (of WHOI) to form a steering group to look into Capart's proposal for the oceanographic laboratory.[49] They thus came up with the plan for a North Atlantic platform, a large manned buoy-vessel combination similar to other facilities (the French Bouée Laboratory and the ONR FLIP—Floating Instrument Platform, for example).[50] The US vessel would reach a specific area in the ocean that the researchers wanted to investigate and then "flip" into a vertical position so as to operate a variety of remote-sensing tools above and below the water surface. The Bouée had recently been used in two subcommittee projects, BOMEX and COMBLAMED 69, which had produced new data on the interaction of air and water masses in the Mediterranean.[51]

Mosby's group underscored the platform's utility to environmental analysis and surveillance alike. Undeterred by the *USS Pueblo* exposé, the

Norwegian oceanographer and his associates emphasized that the vessel could carry on "veille surface" through radar and "veille sous-marine" through sonar.[52] But the platform's cost was well above anything requested to that date by NATO oceanographers.

When Randers received the document, he expressed reservations. The NATO science adviser had become increasingly reluctant to accept the oceanographers' requests, especially in light of the ways in which the first set of CCMS pilot projects were administered. Furthermore, the CCMS attracted a number of marine scientists previously outside NATO sponsorship programs. For instance, those working at Liège's *Centre Belge d'Océanographie* under the guidance of Jacques C. Nihoul had taken responsibility for piloting the CCMS project on coastal water pollution, something that Train openly praised.[53] And Arthur Lee, an ICES-affiliated fishery scientist who had previously quarreled with NIO director George Deacon on NATO's sponsorship of oceanography, had now been appointed chairman of the aforementioned CCMS pilot project.[54] In Randers's eyes, these researchers who carried out their work without asking for additional funds appeared more enthusiastic than their colleagues in the NATO subcommittee.

Randers now appointed the metallurgical engineer Amos J. Shaler to further evaluate the NATO oceanographic platform proposal. Although the consultant viewed the propositions made by Mosby's group favorably and included their recommendations in his final report, the oceanographers complained that his review was imposed upon them. They thus continued debating on the subcommittee's future without even noticing that it was now under review.[55]

Randers's dissatisfaction grew when he examined the oceanographers' management of NATO funds. Capart had recently become the subject of a whispering campaign, and Deacon wrote him that he "sensed growing opposition" to the activities of the subcommittee because he had allowed the development of oceanographic instrumentation of debatable standards. "Someone might grumble," Deacon wrote, about an electromagnetic current meter recently designed by a French firm. He also acknowledged that the subgroup was being "attacked by all sides" for its expensive projects.[56] Capart's various roles in several groups also came under scrutiny for his involvement in the CCMS pilot studies on coastal waters and on sea pollution, as well as for a Science Committee–sponsored conference on Costal Marine Pollution.[57]

The November 1970 CCMS conference on oil spills was not only a catalyst for promoting NATO actions against sea pollution, but also an occasion for witnessing the mounting conflict between groups competing for funds made available by the alliance. In one of the preparatory meetings, brewing troubles became spectacularly apparent thanks to Capart's *faux pas*. His assistant at the Royal Belgian Institute, Elisabeth Peeters, was instructed to send a telex to Shaler asking to postpone the meeting because Capart, as the Belgian representative, could not attend. But Randers informed the participants that a member of the Liège laboratory was present. The meeting should go ahead. At the end of the meeting, Lee reported back to Britain that he felt that the difference of opinion between Belgian groups originated in the belief that Capart had previously participated mainly in an attempt to get "NATO money for his own Institute" and had thus not worked in coordination with his colleagues in Liège.[58]

Capart's circumstances made it difficult for the oceanographic subcommittee to have an authoritative voice. Randers still wanted to empower it and viewed funding for a new air-sea interaction program as the way forward.[59] But that NATO oceanographers were divided over Mosby's project for a North Atlantic platform made him anxious.[60] With the oceanographers not speaking and their chairman ostracized, the chances for the subcommittee's survival were getting slimmer, especially because UK and French delegations were still opposed to a major investment.

In 1972, Randers took advantage of the ongoing restructuring process advocated by the Science Committee to dissolve those subgroups whose finance and research agenda no longer fit within their terms of reference. The Pestel Working Group had by then concluded that new rules for the appointment of subgroups should be defined. Science Committee members would now participate in their meetings and the subgroups would not operate for more than three years without a mandate renewal.[61] This solution pleased French and British delegations, eager to reduce the costs of NATO's science program, as well as US and West Germans, eager to orient it more toward environmental protection.

Randers used the principles set out by Pestel against the advisory group on meteorology. Since the problematic review of 1965, the group had made little progress. If the main problem of the NATO-sponsored oceanographers had been their extravagant use of financial resources, then that of the meteorologists had been their reluctance to engage and spend much of anything. At its meeting a year earlier, Randers had satisfactorily noted that four members (including van Mieghem, Fea, and Flohn) had now retired and that new members were about to be appointed.[62] But by 1970 the change of membership had yielded little. The Science Committee now recommended that oceanographers and meteorologists join forces in a new group focusing on the bourgeoning field of air-sea interaction (examining the boundary layer between atmosphere and oceans that shapes air and sea waves and defines the transfer of energy between the two). Randers, Nierenberg (who by then regularly alternated with Rabi as US representative), Cottrell, and Pestel viewed this proposition favorably.[63] And when the proposal for a merger was finally approved by the Science Committee, the meteorologists endorsed it.[64]

But the opposition of the oceanographers to the changes that Randers advocated was vehement. So on November 13, 1972, Randers attended the meeting of the subcommittee on oceanographic research and announced to its members that the group was going to be discontinued at the end of 1973. From January, an exploratory panel would consider future NATO activities in the field of oceanography. The changes were "noted with regret" by the subcommittee members, who decided not to discuss the other items in the agenda in protest.[65] Although the summary record of the 1972 Science Committee meeting reveals little about Randers's motives, Capart's controversial management of funds accelerated the subcommittee's cancelation. Randers had often pointed out that NATO grantees not only lavished funds, but they did not report regularly enough on their expenditures. During one subcommittee meeting, he stressed that no extension of grants would be forthcoming unless reports were properly filed.[66]

The reproach may well have been put forward for a reason. And Capart and his assistant Elisabeth Peeters seemed once again at the center of Randers's storm. As they had regularly applied for NATO grants in 1971 and 1972, their scientific achievements and management procedures now went under much closer scrutiny. When Peeters applied for an extension of her 1971 grant, Deacon asked the British scientist Alan Preston about her reputation in marine pollution studies. By then, Preston was a renowned world expert in pollution problems, especially in how seaweed harvests dangerous chemicals and radioactive substances.[67] The British expert was unimpressed with Peeters's achievements and anxious about her entrance into pollution studies. "I have no desire to be involved in work with Dr. Peeters," he remarked, having seen his name in the application.[68]

Randers now ordered an additional review as his *coup de théâtre* before leaving office. He asked the US marine chemist Edward D. Goldberg (of Scripps Oceanographic Laboratory, San Diego, California) to carry it out.[69] As US delegate at the 1970 NAC meeting when the divergence of opinion between the Liège and Brussels laboratories had first emerged, Goldberg had taken NATO activities forward by organizing a conference in Aviemore (Scotland) on North Sea pollution. But he did not seem to hold Peeters's research in high regard either,[70] and he negatively reviewed her project. This infuriated Capart, who wrote Goldberg that since he had previously worked at the institute, he "[. . .] had the possibility to address me directly your eventual criticism on our activities. I wonder of your attitude."[71] It is no surprising that Peeters was not awarded a grant, and when a new panel devoted to marine sciences was set up, Capart was not invited to be a member. And when the panel met for the first time, his former colleagues' first decision was to reject Capart's grant application for funding the publication of Peeters's outstanding technical reviews.[72]

The dissolution of the oceanography subcommittee (and the advisory group on meteorology) signaled the end of a defense-oriented NATO science program, but it did not pave the way to promptly embracing one oriented toward environmental protection. Although a group devoted to ecological studies was already active and two more panels would soon be created, their activities had to adjust to the new financial circumstances defined by the economic crisis of the early 1970s. Randers had worked hard to ensure that the Science Committee would be let free to draw a new research agenda rather than be constrained in its planning by financial arrangements with everlasting subgroups. But the circumstances of the global economy undermined his successor.

Economic and Environmental Crises

The Turkish engineer Nimet Özdas took responsibility for steering the committee through the economic downturn of the early 1970s (see figure 5.4). A former student at Imperial College London and MIT, in 1961 Özdas had returned to his home country and found employment at the Technical University of Istanbul. He also took part in key Science Committee meetings, including those typified by the debate on the technology gap. Appointed to chair the Science Committee in 1973, he set himself the task of implementing what the Pestel Working Group had agreed upon in relation to reforming NATO's science program.

But he had to elaborate an emergency scheme first. In October 1973, the Arab Petroleum Exporting Countries declared an oil embargo in response to the Yom Kippur War. The increase in crude oil prices per barrel, which quadrupled by March 1974, prompted a worldwide economic crisis. When it ended, the oil price slowly went down again. By then, however, West Germany's largest private bank had collapsed, and unemployment had risen everywhere in Western Europe.[73]

NATO's science program was badly hit by the spiraling costs. Between November 1972 and 1973, inflation more than doubled in the United States (from 3.7 to 8.3 percent). It nearly tripled in 1974 (12.1 percent) to return to the 1973 rates only in 1976. Western Europe was affected too, and NATO's

FIGURE 5.4. Nimet Özdas (third from right) introducing Advanced Study Institute publications to colleagues at NATO headquarters in 1975. Source: NATO Archives.

Scientific Affairs Division had now to administer the science program in the knowledge that awards could no longer cover research costs in full. Since French and British delegations had not changed their minds about increasing the budget available for research grants, the whole program was on a slippery slope. The budget available increased only by 30 percent between 1969 and 1978. The average size of grants fell by about 50 percent.[74]

In 1974, Özdas ordered the assigning of smaller grants in order to retain a significant number of awards, but one NATO official feared that their size would be "too small to be significant."[75] By 1974, the advisory research grants panel agreed on a policy of austerity: each grantee could have only one award per year.[76] Meanwhile, the Science Committee requested that the NAC give urgent approval of a budget increase to keep pace with inflation.[77]

Financial restraints delayed the definition of new program panels that, in line with the Pestel Working Group recommendations, should have replaced the disbanded oceanography and meteorology subgroups. In 1973, an ad hoc exploratory panel on oceanography headed by the British oceanographer Henry Charnock reported on the future of NATO's oceanography. A meteorologist by training, Charnock had worked at the

SACLANT center before taking responsibility for the rebranded NIO: the Institute of Oceanographic Sciences.[78] Charnock's panel advocated the establishment of two other panels: one on air-sea interaction and the other on the "properties, processes and populations" of the oceans.[79] The latter would look especially into the distribution of marine organisms and the modeling of marine ecosystems with a view to understanding human impacts on marine environments. Goldberg was appointed chairman of the marine sciences panel.

The economic climate forced the new panel to limit its early activities. It eventually managed to secure sufficient funds to promote a program of studies focusing on sea pollution. Goldberg and his colleagues also manifested an interest in promoting synergies with other international organizations. The 1971 NATO North Sea Pollution conference (Aviemore, Scotland) paved the way to a study of the biological processes of the North Sea, which further developed thanks to contributions from ICES-affiliated scientists.[80]

The oceanographic instrumentation put together in previous defense-related research was now made available to NATO-sponsored experts busy charting biological processes and the effects of pollutants.[81] The Aanderaa current meters originally utilized in Cold War oceanographic research now featured prominently in yearly environmental monitoring exercises such as the Joint North Sea Data Acquisition Project. The exercise aimed to find out about the physical environments suitable for plankton's bloom because of plankton's importance for undersea life and marine food chains.[82]

Traditional monitoring techniques proved particularly handy when a ship accident that occurred in the Mediterranean helped Goldberg to demonstrate that NATO's marine science studies could successfully inform environmental policy. On July 14, 1974, a Yugoslav cargo ship, the *Cavtat*, sank off Otranto Cape in the Adriatic Sea. It carried approximately 325 tons of noxious fuel additives (tetraethyl lead and tetramethyllead).[83] The wreck started leaking the toxic chemicals three kilometers southwest of the cape, and nobody knew the risks for humans and marine populations.

The study of *Cavtat*'s pollutants was meant to dispel fears about its nature and effects. The monitoring operations carried out by Italian authorities had initially downplayed the dangers associated with the accident. The collection of sediments and seawater samples revealed no threat, and in 1975 Yugoslav researchers carried out a submarine survey concluding that the pollutants would cause no harm. But then the incident attracted more attention. The renowned French explorer and conservationist Jacques-Yves Cousteau had publicly denounced the risks associated with the ship's poisonous payload. A local Italian magistrate had also ordered the recovery of the drums, because the Italian government was reluctant to start a rescue operation.[84] Nongovernmental organizations in Italy also had been vocal.

The Italian marine biologist and geneticist Bruno Battaglia, a member of the new NATO marine sciences panel and the head of a small hydrobiological laboratory in Venice, which he transformed into the CNR Marine Biological Institute, alerted Goldberg about the controversy now causing a sensation.[85] In turn, the panel's executive NATO officer Tom D. Allan informed the US geochemist Clair Cameron Patterson of the California Institute of Technology and asked for assistance.[86] Patterson's work, not unlike that of Goldberg, had focused on pollutants and, from 1965, he had campaigned against the use of lead compounds. He had previously worked at the University of Chicago, but when he moved to California in the early 1950s he became increasingly concerned about lead levels in the environment, especially after examining concentrations of tetraethyl lead in ice core samples from Greenland. In 1965, he published the article "Contaminated and Natural Lead Environments of Man," which earned him the enmity of experts working for the lead industry.[87]

Patterson was also familiar with NATO's recent turn to environmental research, having taken part in a workshop on eco-toxicity of heavy metal and organic halogen compounds organized by the eco-sciences panel. By the time he was called in to assess the environmental damage caused by the *Cavtat* disaster, Patterson had found that soluble lead could interact with biological matter and be absorbed in marine species, especially phytoplankton.[88] His study was among the first to show that pollutants entering the food chain could cause great harm.

In 1976, the NATO marine sciences panel organized an emergency meeting at Battaglia's institute in Venice. After the meeting, Goldberg submitted a report based on a monitoring exercise and projections of the emissions of lead alkyls from the drums. Data amassed in previous NATO work on water currents was now reused to show that leakage from drums presented no real danger for humans. Yet the drums could burst, increasing exponentially the concentration of lead in water and allowing the poisonous substances released to enter the food chain. If this was going to be the case, consequences would be more difficult to predict.[89]

More detailed work blending chemical, physical, and biological analyses followed. Goldberg also succeeded in gaining the support of researchers in two Italian research laboratories, the Higher Health Institute of Rome and the Water Research Center of Bari, to investigate emission rates.[90] Patterson assisted them throughout. The Italian researchers now concluded that the knowledge on the environmental consequences of moving chemicals across the oceans was inadequate in light of the uncertainty of what could happen when they enter into the biological cycles.[91]

This complexity notwithstanding, Özdas quickly capitalized on the study, which granted to NATO the publicity that previous activities of the oceanography subcommittee had never obtained. On March 10, 1977, an article in the *New York Times* documented the research conducted by Gold-berg's panel members, and the piece was highly appreciative.[92] So in June the Italian government agreed to take action and salvage the nine hundred drums containing tetraethyl lead. The following April, their recovery was completed, thus bringing the environmental controversy to an end.[93] It was never divulged, however, that the measure was preventative and taken in consideration of the absence of final evidence about the consequences of lead poisoning rather than because of conclusive findings. It was the uncertainty of what might have happened that dictated the rescue effort.

That said, Goldberg's panel members had succeeded in blending traditional oceanographic techniques with novel environmental analyses, an ambition that they shared with the other panel assembled after the disbanding of the NATO oceanographic subcommittee. The air-sea interaction panel inherited from the subcommittee the ambition to investigate oceanographic phenomena, but now in conjunction with meteorological processes and especially in relation to the exchange of energy between air and water masses.

Like the marine sciences panel, the panel on air-sea interaction could not begin research activities because of the economic constraints under which it initially operated. In January 1973, its members met for the first time, but its budget allocation for 1974 (87,000 US dollars) was insufficient to organize oceanographic explorations, equip ships with sophisticated instrumentation, and set up modeling platforms. More worrying was the fact that the panel's new chairman, the British meteorologist John Sawyer (who had replaced Sutcliffe as chief research officer at the UK Meteorology Office), continued to show a lack of enthusiasm for the panel's program.[94] Sawyer had already manifested some anxiety about the possibility that oceanographers would take over in the new NATO set up, even if the six-member panel consisted of three oceanographers and three meteorologists.

Notwithstanding financial difficulties and skepticism, the panel brought to completion the Joint North Sea Wave Project, in which measurements of wave spectra were taken through a variety of sensors installed on ships, buoys, and aircraft off the coast of the island of Sylt (Denmark). The oceanographer Karl Hesselmann of Hamburg's *Institut für Meerskunde* prepared the final report.[95]

Sawyer continued to be unimpressed, however, as he let those at home know, and eventually he left the panel, dissatisfied about the influence that the oceanographers exercised.[96] The UK Department of Science and Education (DSE) officials now agreed, in light of Sawyer's lack of enthusiasm, to reconsider supporting a reformed NATO science program. It is not surprising that they reiterated their opposition to increases to the budget available for research grants and rehashed the old mantra that funds should be used only as seed money.[97]

Environmentalism from Above

The establishment of NATO's third dimension gave remarkable opportunities to a number of researchers from its member states to further environmental studies. New research areas (ecology, entomology, studies on air and sea pollutants, eco-toxicology) previously overlooked in the alliance's initiatives now featured more.

This promotion was important for NATO's own diplomatic activities too. The chief instigators of NATO's environmental turn hoped to gain from it in terms of building consensus and promoting new synergies. On the contrary, when the CCMS projects promised to shake up international legislation in the name of environmental protection, allies parted ways again in order to protect national economic interests. Some in particular, like those in Britain, were prepared to embrace environmentalism only when it would not directly affect key projects of its shipping and aerospace industries. And because by the mid-1970s the CCMS had failed to generate sufficient consensus within NATO, US leaders started viewing it as ineffective for strengthening the alliance. Although support for the new committee was not discontinued, its position waned.

The transformative effects of NATO's greening could be felt far more in the context of the Science Committee. The environmental turn encouraged those who sought to use science as a diplomatic device to restore consensus within the committee through the adoption of new programs promoting environmental research. The enmeshing of environmental and science diplomacy through the creation of the CCMS and the committee's reformation produced new synergies between the US and West German delegations. To some extent, the reformation of NATO's science program in the name of environmentalism pleased French and British delegations too, in that it allowed reducing the influence that the oceanographers played in the program without increasing the budget available for NATO studies. And because the French had by now left the alliance's military command, their support to NATO's science and environmental actions could only be seen, especially in the United States, as a decisive factor in strengthening the alliance as a multilateral political entity. The opportunistic nature of NATO's investment in environmental research and actions was plain to see with the economic crisis of the early 1970s: which, if on the one hand it compromised reformation efforts, on the other it provided a justification for downsizing and cutting back on funding. NATO's environmental sentiment was expendable and transitory.

That said, even in the frugality that economic circumstances allowed, the scientists whom NATO sponsored could carry out innovative studies on pollution, such as that on the lead additives that sank together with the *Cavtat* in the Mediterranean. These studies demonstrated that the alliance's grants program could support novel lines of inquiry by reusing knowledge and techniques from defense-oriented research. The next chapter will show that, even in the context of the committee's reformed agenda, in the late 1970s and early 1980s NATO's programs enmeshed environmental and surveillance imperatives in interesting ways.

On the whole, NATO studies played an equally significant role in deflating mounting tensions over existing practices of environmental management and in displaying a naive optimism about the power of science and technology to address the environmental issues of developed countries. Although the *Cavtat* payload was successfully salvaged after the NATO report, the actual understanding of pollution propagation processes was not as clear as the rescue suggested. As Andrew Jamison has shown, the social movements of the 1970s consolidated around groups that advocated a rethinking of the knowledge production process as a whole,[98] whereas not only did NATO's patronage address its own diplomacy shortcomings, but in an exemplary way also countered the radical critiques.

So, when in 1977 the Danish Science Committee representative, nuclear engineer Povl L. Ølgaard, called attention to the "the approaching crisis of science" and recalled that "a distinct movement against technology which had found its expression in the debate on such matters as environmental protection, nuclear power and so on," Nierenberg pounced on him. Pressure groups attacking technology in the name of the environment were only "a vociferous minority." "Most people felt that science and technology should progress without obstruction and with full support," he curtly remarked.[99]

SIX

Strange Bedfellows: Environmental Monitoring and Surveillance

I spent some time in Bavaria where there was a big air control center near Munich. [. . .] NATO had planned a very large exercise and the weather was absolutely horrendous, and so at the very last minute the exercise was canceled. Apparently Soviet espionage had detected the exercise but not the cancelation so on the radar one could see a large number of Soviet aircraft approaching. The crisis was deriving from the question of whether this was a real thing or a normal counterexercise to our canceled exercise. Our commanders took no chances, so full alert. Near four hundred aircraft were soon in the air—ours and theirs. Cold war![1]

The fight against pollution that NATO promoted throughout the 1970s propelled new approaches to environmental monitoring. It also encouraged the search for new ways to confront security threats, blurring the distinction between what the introduction has metaphorically referred to as NATO's "green" and "khaki" dimensions (or, out of metaphor, the alliance's environmental and defense ambitions).[2] This happened at a crucial junction in the history of the Cold War, a point in time when the search for advanced weapons systems made both superpowers more reliant upon scientific and technological innovation. NATO's commanders now sought to find new means to identify and prevent nuclear and conventional attacks, and their Soviet counterparts could now threaten to use faster and more effective weapon delivery systems, which meant less time available for detection, communication, and retaliation operations.

Countering these new threats called for an increase in the automation of NATO's defense systems and defined the war theater as one heavily dependent upon complex electronic systems comprising detection devices, signal routing equipment, and mainframe computers. As Paul Edwards has aptly noticed, the distinctive feature of these new "electronic environments" was an overreliance on man-machine integration, something that—among other things—considerably increased the risk of "trip-wire" conflict resulting from accidental engagement with the enemy.[3]

This reliance also meant that environmental surveillance mattered even more to NATO planners. Signals traveling in water, air, and space became a constitutive element in the accurate calculation of specific responses to incoming threats, and an element that could either prevent or cause conflict in the new electronic battlefield. As the epigraph to this chapter shows, even a simple weather storm had the power to significantly affect detection and response operations in truly unforeseeable ways. As radio signals traveled into stormy airs, they warped and bent, could be transported miles afar and be intercepted, or convey a different meaning. The possibility of conflict no longer rested with decision makers, but with the vagaries of natural events tampering with the strict logic supposedly governing the integration of men and machines. In light of these developments, it is no surprise that the 1983 NATO exercise *Able Archer* was mistaken by the Soviets as a real attack.

NATO's quest for upgrading defense systems with state-of-the-art technologies informed its science and environmental programs' trajectories in two important ways. First, from the mid-1970s, those scientists who had struggled to find a new home in the Science Committee's sponsored environmental research programs ended up gravitating more toward the alliance's defense research organizations responsible for enhancing some of these systems. And, somewhat more surprising, in the running of novel environmental research, one of the Science Committee's subgroups attracted the attention of defense research patrons, thus creating the circumstances for an unusual collaborations between NATO-sponsored scientists interested in environmental protection and those working on defense research tasks. Moreover, while reconfigured as the new object of research-driven policy actions aiming to tackle degradation, the "environment" (at least in the NATO context) continued to be perceived as the traditional military planning trope. It is telling that the alliance's new early warning system was named NATO Air Defense Ground Environment (NADGE).

In retrospect, the synergies between environmental monitoring and surveillance work appear understandable. This book has shown so far that NATO's environmentalism, at least within the research tradition promoted by the alliance, had a distinctive progeny in traditional Cold War surveillance plans. Exactly for this reason, it did not displease the researchers called in to complete surveillance-related work either, in that it further reinforced remote-sensing approaches in a new era heralding its coming of age. But the prospect of interactions with defense research groups caught the members of the Science Committee off guard, because the separation of civilian and military studies sponsored by the alliance had been sanctioned ten years earlier with the committee's crisis. Carrying out joint work with defense research organizations now represented a conundrum to those scientific advisers who had originally orchestrated the alliance's greening because it ran counter to the reconfiguration of NATO science.

Back to Defense Research

If, on the one hand, inflation slowed down the reform of the Science Committee, it also accelerated the dismissal of groups, such as the oceanography subcommittee, whose research trajectory did not fall in line with the Science Committee's new environmental agenda. Some of these scientists now looked for new sponsors within NATO's defense research organizations. From the mid-1960s the Defence Research Group (DRG) started operating as a dynamic unit under the leadership of Norwegian ionospheric researcher and AGARD scientist Finn Lied (see figure 6.1).[4]

Lied organized new research panels responsible for analyzing specific issues under the DRG aegis. The first, on long-term scientific studies, advanced the task originally assigned to the Von Kármán Committee back in 1961. The remaining six panels covered defense areas of interest to the alliance and critical to its surveillance operations (physics and electronics, far infrared, identification of submarines, identification of friends and foes, and land-based air defense).[5]

Some of the Science Committee's sponsored scientists now agreed to work under the DRG aegis, especially when their subgroups struggled to find a role in the reformed committee's structure. For instance, although the Human Factors Advisory Group was restructured as the Science Committee's program panel from the mid-1970s, the group moved closer to the DRG, especially in light of the growing significance of human-machine integration in defense operations.[6]

NATO's changing sponsorship agenda also informed the decision to provide no more grant money to the Joint Satellite Studies Group responsible for NATO's *Spacetrack*. For several years, the group had carried out pioneering research on how satellite signals traveled through the atmosphere—also

FIGURE 6.1. Finn Lied at a meeting of the Science Committee. Source: NATO Archives.

with a view of amassing surveillance data. But the dispute over whether the Science Committee should take responsibility for defense-oriented research affected the group's sponsorship and made it an obvious target for cancelation. At one point, the group's recognized leader, the US radio physicist Jules Aarons, urged Rabi to lobby for the group's survival because it represented for NATO a "best buy" enabling "the operation of a quarter of a million dollar annual program at an annual cost of only $12–14k."[7]

Since Sputnik's times, *Spacetrack* had expanded considerably, especially through the setting up of a network of ground stations detecting signals from beacon satellites across the globe. Aaron's institution, the AFCRC (now renamed as Air Force Cambridge Research Laboratories, AFCRL), had equipped receiving stations at various latitudes from the polar cap (Thule, Greenland) to the tropics (Kingston, Jamaica) with state-of-the-art disk antennae. The data collected had revealed patterns of ionospheric disturbance (or "scintillation") that marred telecommunications, thus further extending the group's research activities.[8] In the running of NATO-sponsored meetings, the group's members reported on patterns of atmospheric and ionospheric absorption at various places through the examination of the total electron content.[9] At this point in time, Aarons had also extended the international collaborative exercise to organizations in non-NATO countries and made available unclassified information (while collecting precious data) in India, Peru, Kenya, Ghana, Israel, Spain, and Jamaica.[10] Because of the Science Committee's reluctance to extend the funding of *Spacetrack*, Aarons renamed the group as the Beacon Satellite Group and looked for a new sponsor, which he eventually found in the International Union of Radio Science.[11]

Studies on atmospheric radio propagation continued, however, to be a distinctive feature of NATO's defense research program, and Aaron retained a role as key contributor in the context of AGARD-sponsored initiatives. This happened at a crucial time, for, by the early 1970s, earth-orbiting satellites allowed the first systematic exploration of how radio signals are channeled (or ducted) into the ionosphere. That signals could be transported far away from transmitters or suddenly fade in proximity to receivers was already known. But the actual analysis of the relevant physical phenomena occurring in the upper strata of the atmosphere was yet to be carried out and, although originally researched in connection with their surveillance, satellites emerged as ideal instruments for acquiring new knowledge on *ducting*.[12] After the launch of AFCRL's Orbiting Radio Beacon Ionospheric Satellite (ORBIS), Aaron and his colleagues explored the distinctive ducting of its signals: they reached the AFCRL Sagamore Hill observatory from the northwest in early daylight and from the south at midnight. With the assistance of mainframe computers and other equipment, data on signal tracks were analyzed and, eventually, displayed the relevant patterns, also giving rise to new hypotheses and theories.[13] Aaron's Italian collaborator Nello Carrara now explained in greater depth how signals bounced off different layers of the atmosphere and were channeled before reaching the earth.[14]

Although the Science Committee agreed that *Spacetrack* should no longer be funded by NATO, Aarons and Carrara's findings on ducting prompted the committee's military authorities to further investigate these complex ionospheric disturbances, because the alliance's defense was wholly reliant on the transmission of radio signals between distant locations. Furthermore, new evidence on ducting in the upper strata of the atmosphere raised questions about the possibility of ducting in the lower strata, especially as a result of meteorological phenomena (such as fog, for instance). Could ducting prevent an efficient transmission or reception of radio signals? Could it disturb the functioning of radar equipment?

Finding an answer to the questions mattered even more to NATO's authorities because, by this point in time, the plans for an independent nuclear force had been mothballed, whereas in the early 1970s NAC gave approval to the establishment of a new early warning system, called NADGE in acronym. This was one of NATO's most impressive and expensive defense project. By 1971, a NADGE management and policy board was operational and, soon after, the board ordered the building of new radar stations equipped with transmission, reception, and tracking devices. Each of these stations covered a specific geographical and airspace area, and their coordination extended coverage to the whole of Western Europe and far beyond. NADGE's core structure entailed an investment of 110 million US dollars and, by 1973, its board had oversight of eighty-four radar stations scattered across Europe and located between North Cape in Norway and eastern Turkey. Three years later, a NATO radar improvement plan also was approved in order to ensure better coverage of the Mediterranean, adding more stations to the existing network.[15]

A consortium of manufacturers, NADGECO, supplied standard tri-dimensional radars (Telefunken, West Germany), height-finding radars (Marconi, United Kingdom), gap-filler radars (Signalapparaten, Netherlands), consoles (Selenia, Italy), and high-speed computers (Hughes Aircraft, United States). Designed to acquire data on incoming aircraft (altitude, speed, course, and identification as friend or foe) and automatically select the most suitable weapon (aircraft or missile), NADGE reached far beyond the Iron Curtain and extended over three thousand miles of airspace from north (the East-West German border) to south (Turkey). Six thousand operators and more than two thousand programmers and other personnel operated the impressive electronic defense system.[16]

NATO authorities justified the decision to build NADGE by citing the recent advances in Soviet aeronautics, and especially the development of the supersonic bomber Tupolev Tu-22. They concluded that, because of its speed, the bomber would go undetected when entering Western Europe's airspace. They aptly code-named the bomber "blinder" and agreed that the new NADGE would counter the new threat by triggering an automated response connected to the tracking system. Such an investment was also rooted in the growing uncertainty that typified Cold War negotiations between superpowers. By the mid-1970s, the Soviet Union had achieved nuclear parity with the United States, also pushing for outnumbering their rivals in terms of nuclear warheads. Key treaties hampering nuclear proliferation, such as the SALT I, led to a less than successful agreement that banned only antiballistic missile systems.

Because of that, both AGARD and the DRG were given much latitude in sponsoring further research on ducting. And so the number of NATO research groups investigating radio propagation now markedly increased. The AGARD Ionospheric Research Committee, as we have seen, had originally pioneered atmospheric propagation studies at NATO back in the late 1950s and had since continued to promote research meetings without, however, receiving NATO sponsorship for original research. AGARD now succeeded in enabling the group to be transformed into an independent panel on electromagnetic wave propagation. Its terms of reference reiterated the focus on propagation problems "not only in the ionosphere, but also in the troposphere, space, the ground and the sea," but it now allowed an investment in original research.[17] Thus, although Aarons took the *Spacetrack* project out of NATO, he also, from 1973, agreed to contribute to the panel's activities. And in 1976 he chaired it, propelling the investigation on beacons and ducting phenomena.

Meanwhile, some of the members of the Science Committee's advisory group on radio meteorology agreed to join one of the DRG's research panels, thus continuing to operate as a cohort promoting an analysis of environmental factors, which had implications for NATO surveillance activities. Actually, the advisory group members shrewdly recognized new research priorities that would invite new sponsors, especially the study of ducting phenomena. Already by the mid-1960s, the members of the radio meteorology advisory group sensed that the Science Committee had become unwilling to sponsor new projects. They thus looked for new areas for investigation and foresaw, not differently from Aarons, that research on atmospheric ducts close to the sea surface could gain them new patrons. Their chairman, Karl Brocks of the Geophysical (later Meteorological) Institute of Hamburg, had envisaged signal ducting to be of particular interest because the sudden variation in temperature between strata produced ducts that affect signal transmission between NATO defense units. Brocks had become director of the institute in Hamburg five years earlier, when he had provided a significant contribution to one of the IGY's successors, the International Years of the Quiet Sun, and revealed for the first time key meteorological features of the tropical regions.[18] The DRG appeared interested in sponsoring new research on ducting problems after Brocks's involvement and that of Norwegian specialist Dag T. Gjessing. He had inherited from other Norwegian scientists that we have previously encountered in this volume, such as Lied and Harang, an interest in using electromagnetic signals as a means of probing the atmosphere. Based at the NDRE station of Kjeller, Gjessing had researched remote-sensing methods extensively while getting more involved in the management of DRG projects.[19]

After NATO's advisory group on radio meteorology was discontinued, Aarons's new AGARD panel agreed to take over some of the group's research tasks (and members). Gjessing and other specialists working on atmospheric ducting agreed instead to join the DRG research unit RSG.6 working under the aegis of its physics and electronics panel. Brocks's successor as head of the Institute of Meteorology in Hamburg, Helmut Jeske, was now appointed to chair that panel and was called in to advance studies on how atmospheric ducting affected the performance of the newly installed NADGE radars, especially in weather conditions such as anti-cyclonic weather.[20]

In launching the new research program focusing on ducting and sea evaporation phenomena, Jeske pointed out that, although the reasons for the formation of a duct over the sea were known, its effects on the propagation of signals were poorly understood and the design of new radar antennae depended on this knowledge.[21] Ducting in connection with radio signals was particularly important in certain regions of the alliance, such as the Mediterranean, where anticyclonic conditions can cause signal drift effects. Signal intelligence personnel had grappled with the problem of environmental signal drift for several years because the location of listening and tracking stations had to be reconfigured accordingly. A US listening station in Ethiopia, for instance, regularly intercepted communications from southern Russia because of that.[22] In the complex game of detections and counterdetections that typified the activities of NATO and Warsaw Pact early warning stations in the European continent, environmental factors could spell success or failure. NATO-sponsored radio meteorologists thus represented a sort of reserve army to expedite research programs that could bring the vagaries of the atmosphere into the systems of accurate prediction and response that military operations required.

Jeske's group was not alone in carrying out environmental studies of interest to NATO's defense establishment. And, although some of these studies were a bit limited in scope and devoted almost exclusively to defense tasks, others increasingly revealed the power of remote sensing. So, during the 1970s, most NATO research groups that aimed at environmental protection and defense were separated by different sponsorship strategies and patrons. But a few of them ended up, unexpectedly, pursuing collaborative work.

NATO's Opaqueness: The Military and Environmental Dimensions of Remote Sensing

By 1978, Gjessing had been busy not only with addressing issues associated with ducting but also with conveying the power of remote-sensing studies to wider audiences. Now appointed chief scientist of the remote-sensing technology program of the Norwegian Council for Scientific and Industrial Research, he completed a monograph in which he emphasized the need for more powerful systems to study the environment and the promise of remote-sensing technologies. In this work, he did not focus exclusively on ducting problems and NATO defense communications but reviewed instead the applications of remote sensing to other fields of human inquiry.[23]

Gjessing believed remote surveillance to be the way forward in investigating environmental pollution and examined in detail remote-sensing methods for the analysis of air and sea pollutants, including chemical agents released in the atmosphere by industrial concerns. He posited using electromagnetic waves as a beacon to reveal the agent's "fingerprints" (specific absorption lines) in the environment.[24] He went on to describe a number of techniques then in use that were based on laser and laser-radar (or lidar) apparatus to detect the presence of carbon and sulfur dioxide and other gases pumped into the atmosphere by industry.[25] He was not alone in emphasizing the power of remote sensing. The European Association of Remote Sensing Laboratories (EARSeL), for instance, used remote-sensing techniques in environmental research, from assessments of natural resources to pollution.[26]

The contiguity between defense and environmental protection research is plain to see, especially in the case of one NATO study, Project OPAQUE, which, while pioneering a technique to improve air defense detection, presented distinctive research features ultimately making the research findings also useful in the areas of environmental monitoring. By the late 1970s, NADGE proved sufficiently reliable in the coverage of the upper and medium strata of the atmosphere but was less so close to the sea surface. Rather than further expand the network, NATO's military authorities agreed to set up the new Air-Borne Early Warning and Control System (AWACS), which consisted of radar equipment installed on a Hawker Siddeley Nimrod reconnaissance aircraft and a Boeing E-3 Sentry.[27] But the accuracy of the surveillance equipment made available for these aircraft depended on the ability of their radars to penetrate cloud formations and their efficiency in clear skies, which instigated more DRG-commissioned research.

Before AWACS became operative, a DRG research group (RSG.8) was tasked to test how atmospheric and weather phenomena influenced the performance of electro-optic sensors such as infrared and laser-based radars. Looking into constituents of the atmosphere, the group agreed to chart how these systems performed under specific weather conditions such as gases, aerosols, fog, clouds, sea spray, and precipitation.[28] Formalized as a project to measure the Optical Atmospheric Quantities in Europe (OPAQUE), it led to a set of airborne and ground measurements of aerosol size and distribution. The project was completed in four years at Aarons's institution (now renamed as Air Force Geophysics Laboratory, Hanscom, MA).

Sophisticated particle-counting equipment mounted on Lockheed

C-130 aircraft produced new environmental data on the optical properties of aerosols at various times of the year, at various geographical locations, and in connection with changing weather conditions. A US Air Force crew flew from more than eight European military airbases in West Germany, Denmark, the Netherlands, France, Britain, and Italy in order to find out more about the "opacity" of detecting devices under various conditions. It is notable that the study was coordinated with research agencies typically devoted to environmental monitoring, such as the West German Meteorological Office and the Visibility Laboratory of the Scripps Institute of Oceanography. The experts of these two institutions agreed to further analyze airborne data collected by the air crews.[29]

The RSG.8 co-pilots (West Germany, France, Italy, the Netherlands, the United Kingdom, and the United States) contributed to the study through the acquisition of ground and sea data. Aerosol and other meteorological parameters were measured with a sea platform (Nordsee) off the Danish coast; the research vessel *Tydesman* of the Royal Netherlands Navy and an RAF research group at Christchurch (UK) contributed to data collections between spring 1976 and summer 1977.[30] In June 1980, a second DRG group under the same panel (RSG.16) was established in order to analyze the data and develop atmospheric optical and infrared modeling.[31]

The exercise highlighted the merits and shortcomings of using coherent infrared radars in low-level reconnaissance exercises with the AWACS system. It also revealed that some detection devices ensured greater cloud penetration but were less reliable in snow or heavy fog. In particular, lidar devices appeared more effective in fog conditions.[32] But although the research carried out by the OPAQUE group was essentially in the realm of defense studies, the techniques adopted during the exercise lent themselves to much more than just improving the use of these systems in defense operations, because they revealed how congenial they could be to an examination of variations in the chemical composition of the atmosphere.

It is telling, for instance, that already in 1975 AGARD had organized a meeting on the optical propagation characteristics of the atmosphere attended by US physicist Gilbert Plass of Johns Hopkins University. By the time Plass was invited, he was known for his theoretical work on infrared radiation in the atmosphere, a contribution that had led him both to explore the effect of infrared absorption in the atmosphere on the functioning of heat-seeking missiles and to consider it as an index of the growing amount of carbon dioxide released in the atmosphere as result of industrial activities, the chief cause of global warming.[33]

Moreover, the role of clouds in radiative transfer through the atmosphere and the use of radar and lidar to measure particulates became subjects of more research on climatic variations. In 1983, the WMO launched the International Satellite Cloud Climatology Project (ISCCP) in the context of its climate research program.[34] And the NATO exercise took place at the same time as leading earth scientists were busy charting the role of atmospheric aerosols in shaping climatic changes. So, while NATO defense scientists involved in the OPAQUE project were apparently concerned with a fairly straightforward defense research task, the subject of their work meshed traditional surveillance work with environmental analyses. We have seen that this was also apparent in the career of Dag Gjessing, one of NATO's funded defense scientists, busy applying remote-sensing techniques to defense problems and eager to apply these techniques to environmental change studies.

After the OPAQUE project ended, the launch of a space satellite equipped with remote-sensing devices allowed NATO to carry out defense- and environment-driven work together. This time, however, the newly established Science Committee's air-sea interaction panel, and not a defense-oriented DRG unit, would call the shots. Actually, in the effort of promoting environmental monitoring studies, the air-sea interaction panel prompted the study of defense-related issues that ended up occupying once again a central role in NATO's science program.

Oil Spills and Their Hidden Military Dimensions

By 1975, just as the inflationary crisis was tailing off, a new generation of meteorologists well versed in oceanographic problems became members of NATO's air-sea interaction panel, including the new panel's chairman, Jan A. Andersen of the Christian Michelsen Institute (Norway). They seemed less inclined than their predecessors to emphasize the problems deriving from mixing groups from different disciplinary backgrounds or research interests as some of their predecessors (i.e., Sawyer) had been. Although the budget available for research grants in 1976 did not increase significantly, the following year new panel members aligned even more air-sea interaction studies to the new NATO environmental agenda by focusing on the impact of this interaction on climatic changes, especially with reference to atmospheric pollution and the movement of jet streams globally.[35]

Andersen and his colleagues showed eagerness to coordinate NATO activities with other international collaborative projects. They eventually joined the WMO/ICSU Global Atmospheric Research Programme (GARP), an exercise aiming to map key atmospheric phenomena in relation to the global circulation of air masses.[36] From 1974, the NATO panel members took responsibility for a large-scale experiment in the eastern Atlantic, the Joint Air Sea Interaction Experiment (JASIN), aiming to chart how air-sea interactions in this sea region affected the circulation of air masses across the globe. The set of experimental analyses covered a sea area near Rockall Trough (off the eastern Hebrides coast) and tested synoptic methods through combined data gathering from buoys, ships, and aircraft.[37]

Drawing on JONSWAP's results, the members of the air-sea interaction panel now envisaged utilizing a satellite in the synoptic JASIN study they intended to complete. This was the period of time when remote sensing (or "surveillance") came of age, revealing the potential for scientific experts to set up complex systems of global monitoring based on sensory devices placed at distance from the objects being studied. It was especially through equipping satellites with a variety of remote-sensing sensors that these experts hoped to secure even greater coverage and accuracy in data collection.[38]

But the potential of synoptic remote sensing attracted other NATO groups too. In 1976, the head of the Department of Pollution Prevention at the French Ministry of Environment, Jean-Marie Massin, asked the French oceanographer Bernard Saint-Guily (one of Lacombe's former assistants), whether the air-sea interaction panel could elaborate new methods to track illegal oil spills. Because Massin was also chairman of a CCMS panel on the use of remote sensing for the control of marine pollution, he considered monitoring one of the stumbling blocks in the adoption of a new oil-spill convention (then being discussed—as previously shown—at IMCO).[39] Massin knew that there was no way to detect spills with the detecting devices available on land, at sea, and in the air. None of them ensured sufficient coverage. And several governments could use the inability of policing the yet-to-be approved convention as an excuse to avoid committing to its approval.

Views on policing changed dramatically after the planned launch of NASA's SeaSat; the first satellite equipped with Synthetic Aperture Radar (SAR) equipment. The new radar allowed the completion of a synoptic analysis of ocean waves, greatly improving both air-sea interaction studies and oil-pollution-policing operations.[40] That a joint exercise uniting research and practical measures to understand and protect the ocean environment could have benefited from the use of SeaSat was now clear to many involved in NATO's science program. But although the plan for partnership between their NATO groups evolved, more experts appeared to be interested. And their roles had nothing to do with oil spills.

Certainly there was some wonder, and possibly some apprehension, among the participants of the air-sea interaction panel when in March 1977 a representative of the US Office of Naval Research, Hans Dolezalek, asked whether he could attend their meeting together with the DRG representative. Dolezalek also agreed to report on the West Coast experiment recently carried out in the Southern California Bight, which had been a sort of trial for equipment to be installed on the SeaSat. SAR (and other sensors) had been mounted on a US navy aircraft in order to gain a better understanding of how satellite-based measurements could be correlated to those obtained at sea by ships and buoys.[41]

Dolezalek's speech was reminiscent of the visit of Captain Kenneth M. Gentry to the Science Committee of twenty-two years earlier,[42] and it resulted in an invitation to replicate the West Coast experiment in the North Sea. The panel's members now agreed to take responsibility for a joint project uniting DRG, the Science Committee, and CCMS subgroups.[43] But though the Science Committee had advocated synergies between the CCMS environmental experts and the scientists assembled in its subgroups for quite some time, the collaboration with NATO defense research authorities was new and was making its members somewhat wary.

NATO's new environmental focus had not reduced the alliance's investment in military and surveillance operations, and this investment took place notwithstanding the mounting conflict between the United States and its allies in Western Europe on coordinated defense and international monitoring activities. A bone of contention was the actual extension of monitoring and verification of atomic tests to international agencies. France had yet to sign or ratify the Non-Proliferation Treaty and was still outside NATO's military command structure.

Once again conversations within the Science Committee weaved in thorny diplomatic issues. The French government had been particularly active, through President Valéry Giscard d'Estaing, in campaigning for an international satellite for monitoring purposes and was canvassing for support among Western European partners. A proposal had been discussed already at the Council of Europe, and led to Giscard d'Estaing's initiative for an International Satellite Monitoring Agency (ISMA) put forward at the UN.[44] The US administration was against it because the French projects would have compromised the ability to unilaterally monitor other countries' nuclear programs.[45]

This is the reason the mounting interest in satellite monitoring of the oceans produced anxiety in Washington. The SeaSat's beacon would be transmitting to the ground tracking network TRANET operated by the US Department of Defense alone. And satellite-mounted sensory devices had promise in a variety of operational applications beyond that of illegal oil spills, including monitoring of shipping traffic and vessels above and below the surface, including nuclear submarines, which presumably US authorities did not wish to share.[46]

The availability of information on the new satellite outside NATO was a particularly concerning item for US officials. By the time the NATO meeting took place, the recently established EARSeL had set up a SeaSat Users Research Group of Europe.[47] The scientist-led organization was backed by the Council of Europe and the European Space Agency. EARSel's chairman was even invited to a meeting of the air-sea interaction panel and had expressed his wish to gain access to SeaSat data to promote work on monitoring satellites in European laboratories.[48]

When the Science Committee met the following November, it became apparent that NATO's involvement had been solicited by the US officials in an attempt to *prevent* other civilian European research groups from gaining uncontrolled access to SeaSat data. So nearly thirty years after its establishment, the Science Committee had once again become a vehicle for hidden defense research ambitions. This time, however, the "call to duty" troubled its members. The renowned geologist John Tuzo Wilson, now the Canadian representative in the committee, stressed that "it was unusual for the Science Committee to cooperate with the Military authorities, and indeed, it might not be entirely wise." An old hand in the room, the Norwegian Finn Lied, recalled that a radio meteorology panel was disbanded to mark a new orientation for NATO-sponsored research. Other science administrators like Rabi and Néel kept quiet, and Özdas replied that everything was being handled with great care. There was a "need to take advantage of the SeaSat data" while keeping a "somewhat low profile approach."[49]

It did help to keep a low profile. SeaSat was operational for only a short time, and this period coincided with the time when the JASIN survey took place. After the JASIN data were put together (between June and October 1978), an abrupt failure of the satellite's power system caused the termination of its mission.[50] The follow-up meetings of the air-sea interaction panel saw the participation of more naval research officers, including Omar Shemdin (of the NASA Jet Propulsion Laboratory), who had designed the SAR device installed on SeaSat. It is telling that this time EARSeL representatives were not invited.[51]

The security implications of satellite remote sensing did not prevent the successful application of US satellites in other NATO-sponsored studies. Actually, in order to keep the initiative's momentum, the members of the air-sea interaction panel now agreed to allocate funds for the organization of a meeting at the University of Hamburg. The oceanographer

Karl Hesselmann suggested replicating the JASIN experiment, and from 1979, two more radar-equipped spacecraft, SeaSat A and Nimbus G, were put into orbit. The panel now launched the project MARSEN (Maritime Remote Sensing) to further refine the analysis of the interaction of winds with the underlying water masses and the generation of waves through the correlation of satellite and surface-based instruments off the Danish coasts (northwest of Helgoland).[52] MARSEN's follow-up, the project NORSEX (Norwegian Remote Sensing Programme), took place in the Barents Sea and Svalbard Island waters. Coordinated by Ola Johannessen of the Geophysical Institute in Bergen, NORSEX was "coordinated but not integrated" with MARSEN as a result of the direct contribution of NASA experts and the security implications of utilizing satellite data.

So, in the age of satellite remote sensing, the air-sea interaction panel got involved in studies that straddled the military-civilian divide. The availability of a new satellite and the wish to test its potential without compromising restricted data informed this contribution. Because the new SAR technology could not be completely framed as civilian, but because the results and data might have possessed military significance, a decision was made to use the NATO panel as a "vector" for channeling data and results. Environmental research and defense imperatives thus overlapped in the organization of a single NATO research project.

This chapter has explained this enmeshing as exemplary of how, even when NATO's science initiatives were reconfigured to accommodate an environmental agenda, its research program did retain features that attracted its military authorities busy planning defense studies. And some of those who had contributed to the NATO program, such as Gjessing, recalled on several occasions the contiguity of defense and environmental applications of remote-sensing technologies. By the end of the 1970s, thus, the "environment" continued to be a construct with a compelling military relevance, even if studies charting it for defense ambitions were now made less visible by the flourishing of new research focusing on environmental degradation. The next chapter returns to the public-facing "third dimension" efforts of NATO, but nevertheless it is worth recalling that its role as science patron in the context of NATO's first dimension (defense) continued to attract more funds than any other sponsorship initiative, especially in light of the renewed Cold War tensions that typified the first half of the 1980s.

SEVEN

Science, Stability, and Climate Change

Some may wonder why NATO would be interested in climate change. To me, this is a bit like asking why a person would be interested in a change in gravity. While gravity does not dictate what you choose to do at any given moment, it does tend to push all your choices in a common direction—down.[1]

In the 1980s, experts and environmentalists alike began to focus more on prominent environmental threats that appeared to have global reach.[2] Sulfur and nitrogen oxides cause acid rains and affect forests far away from the industrial sites that emit them. In 1986, the Chernobyl nuclear disaster produced radioactive plumes that quickly spread over Western and Eastern Europe. The previous year, the discovery of a substantial decrease of ozone in the upper strata of the atmosphere over Antarctica produced anxiety because the "hole" threatened to affect health and environment globally. Meanwhile, environmental scientists had warned that increases in the emission of carbon dioxide caused by industrialization could irreversibly change the world climate. Although they were divided at that time over whether climate change meant global cooling or (as they agree today) warming, they still called for decision makers to act to prevent a future climate catastrophe. In 1992, the UN Conference on Environment and Development held in Rio de Janeiro (Brazil) brought these decade-long concerns to closure by promising new environmental regulations that are still being debated today.

The Science Committee members knew about these global issues too. Yet (notwithstanding the claim of NATO's former secretary general in the epigraph), the historical documentation available suggests that during the 1980s views within the alliance on what to prioritize in its research program differed quite dramatically. So although some committee members advocated focusing on some of these global threats, the program's orientation actually shifted toward other priorities.

The changes occurring in the 1980s further qualify NATO's greening as an effort to align the alliance's sponsorship agenda to the needs of diplomacy, especially as new challenges threatened the stability of this multilateral political organization. The rise to power of Ronald Reagan (US president from 1981 to 1989) and that of the UK Prime Minister Margaret Thatcher (1979–1990) marked the rejection of détente and the return to a doctrine of containment. The new approach to global affairs reignited the Cold War and funneled conspicuous funds toward defense. This stance also posited using scientific patronage as a device to stabilize shaken NATO economies and fragile governments in Western Europe. The alliance now launched the Science for Stability program, which represented a departure from the patronage philosophy that Rabi had advocated in previous years. Contingently, the Science Committee now agreed that the program panels previously set up to implement the alliance's "environmental turn" should be disbanded and that the core of strategic funds previously used for NATO's greening be distributed accordingly to the new stability agenda.

When the program was approved, living conditions had worsened in several NATO countries. Greece, for instance, went through protracted periods of political and social unrest that made NATO an obvious target for demonstrators and therefore gave more power to its opponents in the Western bloc and elsewhere. Stabilization through sponsorship of science initiatives became even more important when the Cold War approached its ending phase, through the fall of the Berlin wall (1989) and the dissolution of the Soviet Union (1991).

But NATO's greening was not over yet. The collapse of the Soviet Empire coincided with the alliance's enlargement eastward, and the dormant CCMS now sprang to life again. Its promoters agreed that it was vital for the alliance get control over, and refurbish, thousands of military installations left rotting in the former Soviet republics and satellite countries. The CCMS thus promoted environmental actions targeting those installations, also recognizing for the first time that military exercises had noxious environmental impacts.

Science for Stability

In the 1980s NATO's science program could consolidate, but it did not expand. It metaphorically resembled a collapsed star whose light kept radiating to the alliance's science and technology community while contracting at the center. At the core of the collapsing mechanism was inflation, whose gravitational pull deprived the program of its energy (i.e., sufficient funds). Lack of increases in funds meant that Science Committee representatives could not think strategically about NATO's investment and assign funds according to an overarching philosophy. Economic circumstances forced them instead to get along by confirming existing strands and promoting ad hoc programs.

It is equally true, however, that the power of inflation was very attractive to those allies who had lost enthusiasm for NATO's science and environmental initiatives. Wanting to reduce the influence of scientific affairs in the alliance's activities, they could now use inflation as a device to reduce the program's impact and visibility without openly challenging its advocates. We have seen that in previous decades refusal to budget increases (especially in the face of rampant inflation) had been the weapon of choice for French and British delegates seeking to disengage and downsize. Moreover, the number of patrons eager to fund research in Western Europe had now increased, visibly shrinking NATO's role as a leading sponsor of research in the old continent. Even before 1980, the Science Committee delegates invited the renowned British physicist Brian Flowers to report on a new organization sponsoring scientific research, the European Science Foundation, that was about to be established.[3] It eventually became a leading patron of science and technology in Western Europe and in the 1990s its investment dwarfed NATO's.

At the beginning of the new decade, Science Committee members sensed that funds ought now to be used sparingly, mainly as seed money to generate catalytic effects (as British officials had often claimed). This meant prioritizing the organization of NATO science meetings and the distribution of fellowships. Grant money—they agreed—should be used more cautiously; primarily for spinoff projects or collaborations that national bodies would eventually agree to sponsor in the future. In 1982, the yearly budget of 5.7 million US dollars was used primarily to fund fellowships (53.5%), thus leaving less than half of the sums available for ASIs (22%). The overall budget for the research grants scheme (comprising individual grants and panel-supported projects) consisted of approximately 1.4 million US dollars, only marginally above the levels of funding of ten years earlier, even though the gnawing effects of inflation had significantly reduced the real value of grants.[4] By then, the science program had contributed to a considerable growth of NATO's research base. Twelve thousand fellowships had been assigned from 1958 to no fewer than seven thousand scholars. Eight hundred ASIs had been completed and sixty thousand scientists attended international meetings. Fifteen hundred NATO projects had made funds available for approximately six thousand scientists.[5]

Could NATO's star continue to shine brightly? The science program had been conceived in the late 1950s in light of contingent factors associated with NATO's own needs for greater integration, the threat of Soviet superiority in science and technology, and the wish to further defense-oriented research. Ten years later, embracing an environmental agenda had successfully extended the life of that program when preexisting research priorities no longer appealed to the alliance's sponsors. But even environmentalism could not make it thrive forever. After NATO's twentieth anniversary, its secretary general Joseph Luns and science adviser Nimet Özdas urged Science Committee delegates to diversify the program. They went on to postulate embracing the human and social sciences because new methods were needed to tackle pollution, energy provision problems, and health issues.[6] Luns claimed that the program would retain an environmental focus and that NATO would tackle new challenges "of a global nature including the protection of the environment, the management of natural resources, the welfare of our peoples."[7]

In reality, exactly because of its budget, NATO could not diversify its science program as Luns and Özdas hoped. The increase in defense expenditures was a factor preventing new initiatives. After the 1981 election of the Republican Ronald Reagan as US president, the budget for defense in NATO countries increased dramatically, affecting funding in other areas. Reagan pursued a policy of containment that took East-West relations back to Eisenhower's tense times. After Soviet leader Leonid Brezhnev's rejection of the "double-track" offer of December 1979, namely to reduce nuclear arsenals in both blocs to avoid a massive military buildup, Reagan ordered the deployment of the intermediate-range Pershing and cruise missiles in Europe.[8]

Brezhnev continued to promote bilateral arms-control measures. But from 1979 on, the Cold War turned hot in Soviet-invaded Afghanistan, and in 1983, Reagan's plans for the Strategic Defense Initiative (or Star Wars) envisaged the possibility of setting up a defense system in space, disrupting disarmament talks too. The US president's advisers went on to reelaborate NATO's strategic posture, and, although McNamara's old flexible response was not yet abandoned, it was recognized that the doctrine should encompass an "active" defense element. The resulting Deep Strike strategy projected a limited intervention of NATO forces in enemy territory in order to further increase flexibility in the event of a conflict. It is not surprising that the strategy further destabilized East-West relations in the unfolding of the Euromissiles crisis.[9] And it increased the chance of what the previous chapter identified as the risk of "tripwire" conflict resulting from an accident, such as in the case of *Able Archer*. A NATO exercise consisting mainly of simulating the readiness of the alliance's forces, it actually nearly produced a nuclear exchange between superpowers in November 1983 as the Soviets misunderstood the exercise as anticipating an attack and readied to retaliate using their nuclear arsenal.[10]

A tenser Cold War meant fewer NATO funds for blue-sky thinking and more for concrete measures that could strengthen the alliance. A small Science Committee group, led by the Dutch representative and president of the Netherlands Organization for Applied Research L. B. J. (Louis) Stuyt, was now assembled to plan the NATO science program's future.[11] But lack of funds and defense imperatives successfully geared it toward the ambitions of fortifying the administrations of struggling NATO countries.

Rabi, who continued alternating with Nierenberg and with Nixon's former science adviser Edward E. David as US Science Committee representative, manifested his country's apprehension for the lack of interest that the allies manifested in NATO's investment in science and technology as such. He desperately reiterated that only by concentrating efforts in a few key areas, including climate change, which could offer returns in terms of impact on the cultural debate, the program would thrive again.[12] But this time Rabi's words failed to persuade. Stuyt's Ad Hoc Group of Expert Advisers on the Future Work of the Science Committee (also including Lied, Alves Martins, the Dane Ølgaard, and Cottrell's successor, the physicist Sam Edwards) filed a report highlighting that this time NATO's lump sum for collaboration was going to be invested in the Science for Stability program.

This assistance scheme targeted less developed NATO countries (Greece, Portugal, Turkey) using the sponsorship of scientific projects as a means of political stabilization and economic growth. It was, to some extent, a product of the "technology gap" notion that a decade earlier had produced a heated debate within the Science Committee and saw Özdas (currently the science adviser) joining forces with the Portuguese Alves Martins and the Italian Giacomini. Özdas might have now supported the new plan for similar reasons, but it was especially the alliance's current circumstances that convinced Stuyt's group members to approve it, for the economic conditions of these countries had deteriorated rapidly and their political stability was affected by the decline. As in previous years, views on the proposal in Washington differed quite dramatically. Rabi and State Department officials did not like the new program because they were traditionally opposed to using science sponsorship as assistance money. But other departments, including the Department of Defense, were presumably more appreciative, especially because NATO was about to enter a new phase of the Cold War that demanded financial vigor as much as military strength.

Scientific patronage was now reimaged as a more rudimentary diplomacy device, used to strengthen frail regimes in key NATO territories (including its outer ring). Greece was a case in point. The country's colonels' military coup failed to produce long-term governance and students of the National Technical University of Athens, home to several NATO projects such as those previously directed by the radio meteorologist Anastassiades, organized protests that led to social unrest and criticism of the alliance's role in Western Europe. Even officials of the Greek navy participating in NATO exercises mutinied, revealing that unrest existed within the Greek armed forces too.[13] Italy and Turkey were administered through more stable (albeit short-lived) governments, but as Daniele Ganser has argued, their stabilization came as a result of a "strategy of tension," namely the support of far-right terrorist groups by covert agencies to weaken protest movements.[14] The Turkish invasion of Cyprus, a crucial outpost in the Mediterranean, escalated a major conflict between two NATO allies, Greece and Turkey, in a strategically vital area. Another military coup took place in Portugal, overthrowing a military regime (the *Estado Novo*) and replacing it with another one worryingly close (from a NATO perspective) to socialist organizations.

These regime changes and conflicts paralleled the mounting social unrest in NATO countries, which was another factor instigating the approval of the new stability program. Science was now seen by protesters as part of the problem of their modern societies rather than the main solution. The deterioration of support to basic science among researchers had resulted, Stuyt's group had concluded, in a "backlash of the anti-technology movement," a movement that had swept across Europe for nearly ten years, producing a critique of the role of science in Western societies.[15] Because of this backlash, the number of research applications to NATO programs had significantly diminished too. During one of the Science Committee meetings, Rabi had nervously confirmed that one recent ASI meeting had "attracted people whose ideas ran counter to NATO and to NATO support." Özdas also pointed out that the "NATO label" discouraged scientists from applying for grants and fellowships.[16] The new Greek representative in the Science Committee, George Contopoulos, was targeted by the students in Athens; graffiti on the city campus read "Down with CoNATOpulos."[17]

The criticism of NATO science and scientists was an integral part of the disapproval of the alliance and its role in world affairs. As Jamie Shea argues, "By the early 1980s NATO was facing an existential crisis" with "hundreds of thousands of people demonstrating on the streets" and "NATO governments at the brink of resignation over the decision to install cruise and Pershing weapons."[18] In 1983, three hundred thousand protesters demonstrated in Berlin before listening to Chancellor Willy Brandt's speech against the new missiles, and the Campaign for Nuclear Disarmament's consensus was growing in Great Britain. Many manifested their disagreement with the progressive renunciation to arms-control initiatives in favor of rearmament, while the 1970s pledge of Pierre Harmel and others in favor of an East-West dialogue seemed to no longer inform NATO's agenda.

Stuyt group's report thus envisaged the need for interventions to defeat criticism in less developed NATO countries[19] and for continuing NATO's investment in scientific research through its chief science organization: "The Committee must continue to be the harbinger in the international arena for proclaiming and promoting the importance of new knowledge and assessing the health of fundamental research."[20] It went on to stress the need for *targeted* interventions in order to stabilize precarious political and social regimes. The special fund thus was meant to improve the conditions of the research base in less developed NATO nations, and enable the purchase (or manufacture) of new research facilities and equipment.

Science for Stability proved long-lasting and evolved in three phases. The first phase was completed in 1987, the second in 1993 and the third in 1997. Fifteen million US dollars were disbursed in the first phase, twenty-five in the second, and a further six million in the third. The program objectives were that scientists in Greece, Portugal, and Turkey could set up projects, build cooperative activities, and manage major research endeavors. Funded projects focused especially on the evaluation of natural resources, new methods of environmental management, the production of new energy systems, and improvements in food manufacturing.[21]

The new NATO projects allowed the synthesis of new materials, the production of equipment (lasers and computers), the designing of therapies against infectious plant diseases, archaeological studies, and much more, but even NATO evaluators judged its results to be not wholly satisfactory: for instance, a half-million-US-dollar investment aimed at restoring degraded forests in Greece. But when the scientist responsible for the project died, his successor diverted the funds toward a molecular biology study on olive cultures, which wasted the funds his predecessor used to purchase equipment that was never used. The NATO steering committee called in to review the project's assigned marks.[22] The aforementioned project was rated zero for most criteria, and only the international contacts that it created were judged satisfactory.[23] It was not the only disappointment. A laser-cutting machine designed to make plastic bags in Portugal was never completed because the investigators "seem not too well versed in the optics that lie [sic] at the core of their project."[24] One Turkish project aimed to make recommendations for the production of eco-friendly designs of traditional coal heaters and successfully accomplished the task in 1988, but by then Ankara was switching to natural gas heaters and the market for coal stoves had flamed out.[25]

Because Science for Stability attracted conspicuous funds, funds for other strands of the NATO science program did not increase significantly. And the panels established when greening the alliance had been a Science Committee priority were now put under review. This did not necessarily mean a shift in emphasis, and the NATO program retained a focus on the production of environmental knowledge, but the committee recognized that the vectors through which this knowledge should be delivered ought to change. The result was hodgepodge, however, and exactly when several new global environmental threats, including global warming, urged investigations, NATO downsized its investment.

The "Volcanic" Beginnings of NATO Climate Change Studies

In the 1980s, environmental degradation had truly become a global phenomenon calling for new regulations across national borders. A number of European studies focusing on acid rains revealed sulfur and nitrogen oxide emissions to have "transboundary" effects, which prompted the Council of Europe to set up a European monitoring program. Earlier initiatives existed too, for the OECD also pioneered and sponsored novel research.[26] Notably, these investigations had shaped transnational alliances as, for example, when Scandinavian countries and Britain initiated collaborations. New environmental diplomacy work emerged from this; in 1979, the Long-Range Transboundary Air Pollution convention was signed at the UN.[27]

The need for transnational environmental legislation appeared even more compelling after the 1984 disaster at a chemical plant owned by

Union Carbide in the Indian city of Bhopal, which revealed that the polluting activities of multinational businesses were harder to scrutinize under mere national regulations. But it was especially the sensation created in 1985 by the findings of a British Antarctic Survey research group (later confirmed by NASA satellite observations) about the growth of a large hole in the ozone strata covering the Antarctic sky that threatened environmental concerns globally. As Karen Litfin has highlighted, these findings propelled environmentalism in the international policy arena by creating the circumstances for approving new regulation globally banning ozone-depleting substances (the chlorofluorocarbons, CFCs)—the so-called Montreal Protocol.[28]

Almost at the same time yet another global environmental threat caused anxiety as scientists were now more vocal about major climatic changes. The scientific community had that far offered mixed message on these changes as the early bleak forecasts of a new ice age were replaced by diametrically opposite predictions of a global warming. Speaking in 1979 at the WMO World Climate Conference, the NATO science veteran Frederik Kenneth Hare had scrupulously weighted evidence in favor of both cooling and warming and concluded the second to be likely.[29]

But NATO's role of the previous decade as promoter of environmental studies now waned somewhat. Its environmental actions now shifted especially to avoid controversial themes, such as nuclear energy, that had instead made grassroots activists more vocal. As Stuyt group's report was approved, Özdas was appointed Turkish Minister of State for Science and Technology and the French physicist Robert Chabbal replaced him. His appointment as a Frenchman to be science adviser in a historical period when France was not integrated in NATO's military command demonstrates how the search for new political synergies could be invigorated by science diplomacy. In particular, the appointment coincided with the realignment of French and US administrations that followed the appointment of Giscard d'Estaing's successor, François Mitterrand, as president of France. It is notable that Mitterrand supported the installation of cruise and Pershing missiles in West Germany notwithstanding the anxiety of his German neighbors.

Chabbal had vast experience in the administration of science policy in international agencies, which he had developed especially within the European community's research organizations.[30] As Denis Guthleben shows, he rose through the ranks in the French Centre National de la Recherche Scientifique (CNRS; and became its director in 1976) by pioneering a new concept of state sponsorship of science emphasizing the socioeconomic impact of novel research and the strengthening of connections with the industry and business (especially in environmental areas with promise such as solar energy).[31]

During Chabbal's reign, major changes in the NATO science program took place, also shaping the alliance's investment in environmental research. For instance, in 1982, the committee approved the Double Jump program aimed at sponsoring research in collaboration with the private sector. This was a new emphasis driven by the Silicon Valley model pioneered in the United States and marked the spreading in Europe of a yuppie culture asserting the merits of cultivating science and technology as money-making enterprises. The Double Jump program (surmounting barriers in the exchange of science and technology between research sectors—the first jump—and countries—the second) led to an investment of about one hundred thousand US dollars each year from 1982 for various initiatives, including the awarding of fellowships, grants, and ASIs.[32] Chabbal did not advocate the end of an environmentally driven NATO research agenda, but he gently pushed toward what Jamison describes as "the commercialization of environmental politics" through the setting up of green businesses and the blending of corporate, governmental, and nongovernmental actors and interests.[33]

Chabbal's administration also coincided with the end of the environmentally driven Science Committee panels set up in the previous decade. He instructed those responsible for the eco-sciences panel to provide a statement on future directions of ecology research, which on its own was deemed insufficient to justify further NATO disbursements. The Science Committee members also urged the panel to consider how to present ecological concepts to planners and decision makers—to develop more effective communication strategies. They eventually agreed to renew the mandate for the panel, but only until 1982, so that its members could launch a new study on the mathematical modeling of ecosystems.[34]

The marine sciences panel was also up for review in 1981. The document on the panel's history prepared on the occasion of the twentieth anniversary acknowledged that national resources available for research had increased considerably, thus making it possible for the panel members to find new sponsors.[35] These opinions stirred a controversy (or what according to the minutes of one of the Marine Science panel meetings was "a lively [nonminuted] discussion"), for some Science Committee members felt that the panel had initiated commendable projects but also felt that the statement on scope was outdated and should be revised.[36] The panel's mandate was extended for two more years, but the committee wanted the panel to look more vigorously for other sponsors.[37] It confirmed support to the yearly budget of about $115,000 per year but agreed that the panel was to be discontinued.

The air-sea interaction panel did not survive its review either. In a document prepared for the committee, its members claimed to have worked with proficiency in the organization of a variety of useful tasks. These included projects to advance meteorological science, to address environmental challenges, and to meet the new demands of fundamental knowledge, especially new methods of monitoring and remote sensing.[38] But the panel's focus on the European environments was now considered too narrow, especially because international organizations such as the UN, the WMO, UNESCO, and ICSU had all launched major programs, like the GARP, seeking global coverage.

During the panel's review, the question of whether NATO should contribute to the study of climate change was openly discussed, for the subject had now traction within the community of environmental scientists. When the members of the air-sea interaction panel met for the last time, they posited that the alliance had already supported GARP. Because the launch of a World Climate Research Program (WCRP) was imminent, the panel should continue to invest in environmental research, possibly with a focus on climate, especially considering its implications for global affairs in the future.[39]

When this call was made, however, Chabbal had already made up his mind not only on canceling the existing panel but also on what kind of environmental studies the alliance would fund. In particular, he believed previous exercises consisting of merging NATO groups (such as those on oceanography and meteorology) to have been successful and thought to revamp NATO's science program through the establishment of a new panel exploring air, sea, *and earth* interactions.

He thus ordered that prominent experts on these interactions meet, which is what happened on October 3, 1981, when the Icelandic geochemist Guðmundur E. Sigvaldason chaired the so-called "geo sciences survey group." He invited two scientists involved in the World Climate Research Programme: the new director of the UK Meteorological Office, John T. Houghton, and the professor of oceanography at the University of Kiel (West Germany), John Woods (also a former member of the air-sea interaction panel).[40]

A professor of atmospheric physics at Oxford University, the Welshman Houghton had encouraged the analysis of major climatic variations, including the *El Niño* effect, an intense band of warm waters that forms in the Pacific Ocean and affects climatic events in America.[41] Houghton had thus contributed to the design of the World Ocean Circulation Experiment. Its tropical equivalent, the Tropical Ocean Global Atmosphere, was instead the brainchild of John Woods.[42] These exercises aimed to pin down the key processes governing climate globally, thus producing the final evidence needed to assess the impact of human activities on climate. When global warming became a matter for international negotiations, both Houghton and Woods continued to be protagonists in these conversations at the boundary between doing science and making policy.[43]

Climate change was thus going to be a key research item in the new NATO geoscience panel's agenda, as the invitation of Houghton and Woods to its planning phase made patently clear. Yet Chabbal's double-jump philosophy meant that he agreed to invite to the meeting other scientists from solid earth research studies. One was the geochemist Guð-mundur and the other was his French colleague Claude Allègre of the Paris-based *Institut de Physique du Globe* (IPG).

To understand the reason these scientists were canvassed, we have to look into the Soufrière Affair, a case in which they and Chabbal had played an important role and had, at the time, been decisive for the reorganization of national research in France. *La Grande Sufrière* is an active volcano on the island of Basse-Terre in Guadeloupe (a French overseas territory in the Caribbean). In 1976, the volcano erupted, raising the issue of whether seventy-three thousand Guadeloupians should be evacuated from nearby areas. They were relocated, eventually. But the seismic and volcanic activity did not justify the decision, because the eruption caused less damage than expected. No lives were lost, but a bitter controversy ensued, due mainly to the Polish-born French volcanologist Haroun Tazieff, head of the IPG volcanology section. He had downplayed the possibility of a major eruption for which an evacuation was needed, but Allègre, who was appointed by Chabbal to assess the situation, decided to recommend evacuating the area anyway. So, in 1976, Tazieff left the IPG, and in a *Nature* article he accused Allègre of practicing *terrorisme intellectual*. The decision to evacuate was "expensive and futile."[44]

The *Journal of Volcanology and Geothermal Research* picked up on the controversy and called for a deontological code for volcanologists.[45] Guð-mundur now defended the IPG and publicly rejected the proposition, also recalling that when the critical set of decisions leading up the evacuation was made, Allègre's envoys, rather than Tazieff, had participated in all the meetings. Guðmundur also highlighted that Chabbal, as CNRS director, had played a positive role in the affair, in that he had defended Allègre for the decision to evacuate.[46] In 1980, Tazieff responded to his critics, but his précis was not timely. In the same year, Mount St. Helens erupted in the state of Washington (United States), and Tazieff flew over the volcano, declaring it not dangerous to the population. An evacuation zone was set up but proved insufficient. Fifty-seven people were killed following the eruption, demonstrating that volcanic processes were difficult to predict even for the fiery Tazieff.

The affair was decisive for Chabbal's career, and he rewarded colleagues who had stood with him by inviting them to the NATO meeting. So when plans for a geosciences survey group were put together, it was not surprising that Chabbal suggested inviting the two experts who supported him in the affair. But the idea of assembling earth scientists and climate change experts had critical flaws. During the meeting at NATO headquarters, lack of communication between scientists interested in different branches of the geosciences emerged. It was not necessarily that they did not agree; one group member simply did not know what others were busy doing in their respective fields.

Moreover, atmospheric and earth scientists had recently clashed over global warming, as volcanologists had been involved in another controversy invested in issues more directly related to climate change issues. During the meeting, the volcanologists attempted to convince Houghton and Woods that carbon dioxide released from the active volcanic rift zones of the earth was far higher than any anthropogenic activity recorded in history.[47] This was partly because another controversy on past climate catastrophes was churning on the other side of the Atlantic. In 1980, the physicist Luis Alvarez had concluded, on the basis of an analysis of iridium deposits, that dinosaurs had been killed by a meteorite hitting the earth. But volcanologists discarded the impact hypothesis because of their understanding that iridium is a substance that can just as well be produced during volcanic eruptions.[48]

The polemic also overlapped the debate on nuclear winter, namely that global nuclear exchange could cause a dramatic shift in global climate and instigate not global warming, but a new ice age. This was a debate of great interest to some of the invited scientists, such as Houghton, but with broader relevance to NATO as a military alliance since the results of novel research on the predicted catastrophe emboldened the ranks of antiwar protesters against the alliance's role in the reignited Cold War.[49]

The possibility of climate catastrophes divided the 1980s scholarly community, and earth scientists such as those that NATO had invited to set up a new program emphasized that the issue of time scale made it difficult for the survey group members to pool together methods and practices adapted to examine environmental phenomena: "while processes in the atmosphere and oceans can be measures on a seasonal timescale, processes in the solid Earth take years, perhaps thousands of years, or millions of years."[50] Responsible for at least jotting down some of these observations, Allègre had previously explored the formation of the atmosphere in the early history of the earth and taken part in NASA's Apollo lunar program.[51] The *impromptu* remarks, however, revealed that the earth scientists in the room were presumably unconvinced about anthropogenic climate change.

So the geoscience research group did not prioritize its study. Three avenues of research were highlighted in the initial meeting between the climate change advocates Houghton and Woods and the skeptics Guðmundur and Allègre: studying the planetary climate system; considering the remote sensing of atmosphere and oceans; and focusing on transport problems on a planetary scale.

Allègre and Guðmundur now advanced the idea of understanding climate change as a consequence of variation in the solar atmosphere to overcome "the older 'weatherstation' approach of the geographically inclined climatologists of the last generation."[52] But Houghton and Woods's counterproposal indirectly emphasized the merits of such an approach: it was a plan to advance prior air-sea interaction exercises (from JASIN to MARSDEN) by using remote-sensing techniques to explore the atmosphere and the ocean. It is not surprising that only a third proposal seemed viable, for it entailed an examination of major geological processes involving the solid earth, the atmosphere, and the hydrosphere.[53] But the conversation between experts unwilling to listen to each other seemed to have ended there and concluded with the proposal of organizing workshops. The first, on the interpretation of remote-sensing data, appealed to the climatologists only, and the second, on the dynamics of magma chambers, was of interest to the earth scientists.

In 1983, the panel was finally assembled and the planned workshops completed. The Science Committee, however, instructed the panel to focus exclusively on the problem of global transport mechanisms in the geosciences, which in fact was the name with which the panel was later identified. The crowding together of research areas as different as the formation and distribution of economically important ore deposits and the mitigation of natural hazards did not help to produce synergies, nor did it resonate with the growing attention on climate change.[54] Houghton soon lost interest in the panel; Woods agreed to be one of its members and Allègre chaired it.

If we take Spencer Weart's remarks about the early 1980s as when a "revolution" took place in the social structure of climate science via the adoption of new scientific tools and more international networking,

NATO did not represent a key node in the network then taking shape partly because of disagreements between its scientists on the causes of climatic change and also because of the implications of this scientific debate in apocalyptic predictions on the outcome of nuclear exchange.[55] Disbanded in 1989, the panel was replaced with a new one on the science of global climate change, putting more emphasis on interdisciplinary research as the panel assembled "physicists, chemists, geologists, climatologists, biologists and ecologists."[56]

By then, however, an Intergovernmental Panel on Climate Change (IPCC) had already been set up and gave to the two organizations cofounding it, the WMO and UNEP, the power to inform, through scientific advice, policy proceedings at the UN. This, albeit in controversial ways, led to the organization, under the UN umbrella, of policy propositions aiming to tackle carbon emissions. Notably, the IPCC appointed Houghton as cochair and Woods led on drafting the first IPCC report.[57] Four years later, Earth Summit in Rio de Janeiro brought the results of scientific inquiry into the international policy arena and propelled the definition, in a nebulous form, of a treaty calling for a drastic reduction in greenhouse gas emissions. The focus of environmentalists and policymakers now had shifted from the generic stance of what quality of life could be expected in a depleted environment to sets of responses to climate change: from mitigation to adaptation and geoengineering. Not just that. Fantastic schemes rivaling Edward Teller's earlier propositions to NATO officials for weather control suddenly reappeared on the international scene and were widely debated by scientists across the planet.[58]

NATO's science administrators who followed Chabbal attempted to regain a prominent role in this debate. From 1991, a NATO special program on global environmental change began, consisting of approximately fifty workshops producing an impressive number of publications with the support of publisher Springer, with whom the NATO Scientific Affairs Division had then agreed to work in partnership. Some of these initiatives saw the participation of leading climate change experts such as the nuclear winter theoretician Paul J. Crutzen, who had by then advanced the idea of a new geological era, the "Anthropocene," characterized by man-made planetary changes. The special program even pioneered studies linking climate change to political and social unrest by looking at the shaping of ancient societies with a view of offering an overview of what the future might bring.[59] But NATO lost an opportunity to have a more authoritative voice in the international debate on climate change so that the environmental groundwork typifying the 1970s "greening" of the alliance bore less fruit than expected.

Actually, an interesting question that future scholars will presumably be able to address with greater accuracy than this book does is whether the forging of new diplomatic science ties at NATO during Chabbal's regime might have been instrumental in strengthening consensus *against the notion of* climate change. Years later, Allègre ended up at the center of a public storm for joining the skeptics' camp and was accused of acting not from his scientific convictions but rather from convenience.[60] And, as Naomi Oreskes and Erik M. Conway have recently shown, at the end of the decade two US science administrators who had gained prominence within the alliance, William Nierenberg and Frederick Seitz, launched a campaign against scientists warning about climate change in the context of the US conservative think tank George C. Marshall Institute. Nierenberg went on to reject the findings of the first IPCC report (1990) insisting that global temperatures would increase by no more than 1°C by the end of the twenty-first century.[61] But Nierenberg still had an important role in the planning of NATO science while doing all this (as both a former NATO science adviser and a former US representative in the Science Committee; see figure 7.1).

Oreskes and Conway also claim that changing patronage circumstances at the end of the decade played an important role in Nierenberg's crusade as the now-approaching end of the Cold War called for rethinking individual and institutional research agendas. In 1989, the fall of the Berlin Wall paved the way to the changes of regime in former Warsaw Pact countries. In 1990, East and West Germany reunited and, the following year, the Baltic republics became independent. Nierenberg's campaigning reaffirmed that it was wrong for government scientists to succumb to the climate change alarmism propounded by the same grassroots protesters he had labeled at NATO headquarters as a vociferous minority.

Nierenberg's skepticism on climate change echoed the inconsistency on environmental matters of the new US president elected to replace Reagan in 1989. Back in the 1970s when Nixon had called George H. W. Bush to be the US ambassador at the UN, the diplomat had wholeheartedly embraced the president's environmental turn, emphasizing that he intended to build a "dynamic environmental program" within the foremost supranational agency. But when elected president himself, Bush took advantage of Nierenberg's inspired polemic to reduce funding for climate change studies. In 1992, Bush confronted statesmen and activists meeting at Rio's Earth Summit, plainly rejecting proposals for a deal on cutting carbon emissions and reiterating in private (allegedly) that the American way of life was not up for negotiation.[62]

Future research will clarify whether climate change occupied a pivotal

FIGURE 7.1. William Nierenberg at a Science Committee meeting in the 1970s. Source: NATO Archives.

role in NATO affairs because of a wish to shy away from studies on the nuclear winter scenario, the emerging post–Cold War challenges, or rather for the opinions of some of its key players, including Nierenberg and Allègre. The first visible consequence of the massive transformation happening in the post–Cold War period was the reduction in the number of NATO divisions stationed in Western Europe. The number of US troops in Europe also fell, from 336,000 in 1990 to 195,000 in 1994.[63]

The alliance was rapidly changing as a consequence of the end of the Cold War, and its aftermath was decisive for both the political and the defense dimensions of the alliance. And although disengaging from environmental research deemed decisive for the future of the planet, NATO continued, in fact, to promote environmental work. This promotion was in the context of activities more central to its political and military missions. The alliance turned "green" once more in an effort to assert its influence in now-defeated regimes of the Eastern Bloc.

Revamping the CCMS

The CCMS had survived the 1974 review and (according to Jacob Hamblin) "soldiered on" in the following decade.[64] During the 1980s, the committee continued to address a variety of environmental issues, especially in an effort to counter what Stuyt's group had characterized as the "anti-technology movement." To some extent, this effort resulted from the fear that this movement could take environmentalism in its stride and move the debate on environmental management toward radical political solutions. More generally, 1980s environmentalism shaped new models of participation in local decision-making, especially in Scandinavian countries.[65] So, for instance, Luns condemned at once mass protests against nuclear energy programs in NATO countries, claiming that "it was imperative that these should be considered on a sound and rational basis" and advocated a number of pilot programs that would allow successful management of environmental issues without requiring major structural changes to energy production processes.[66]

Keeping environmentalism firmly within the realm of a "stabilizing" rhetoric helped the CCMS to launch new initiatives such as the adoption of satellite methods to police illegal oil spills discussed in the previous chapter. In 1985, the committee launched a "dioxin problem" pilot study as a response to two major incidents in which the poisonous chemical had been released in Seveso (Italy, 1976) and Times Beach (Missouri, United States, 1982). NATO now looked for ways to further dampen controversies over dioxin's uses, especially from grassroots protesters.[67] But environmental disasters and the Cold War crises of the Reagan-Bush era lent further support to environmental organizations. For instance, the sinking of a Greenpeace vessel then busy stopping a French nuclear test spectacularly increased public support for the eco-warriors.[68] The CCMS program was not entirely in sync with the spreading mass protests, especially against nuclear energy (the "Revolt of Doves").[69] These protests targeted NATO, recognizing it as responsible for both the Euromissile crisis and for support to civilian nuclear power. In turn, this reduced the CCMS's appeal to the point that by the time the Soviet Empire crumbled, many within the alliance wondered whether the committee still had a reason to exist.

There is currently little evidence in NATO Archives on whether (and how) the 1986 Chernobyl nuclear power plant meltdown changed perceptions within the alliance on the viability of atomic energy. Five years after the disaster, one CCMS booklet emphasized the merits of the air pollution models elaborated by the committee in the early 1970s in making predictions on how radioactive contaminants spread in the atmosphere. In 1991, however, the committee phased out an air pollution modeling program and seemed oblivious to issues such as the environmental impacts of nuclear energy and waste.[70] With the antinuclear movement now at its peak, the committee agreed to investigate and promote actions for improving preparedness. The new pilot project consisted of finding out (two years before the Chernobyl meltdown) how to improve medical coordination in case of future nuclear disasters.[71] So, contrary to the widely held belief that to prevent is better than to cure, the CCMS was eager to find ways to better cure people across Europe in case of more Chernobyl-like incidents but not enough to understand how to prevent them.

A number of CCMS projects, especially in the 1990s, tackled health and environmental issues associated with radioactive contamination, but these actions followed a period when the environmental impacts of nuclear energy had been overlooked. This absence is even more noticeable if one considers that the Chernobyl accident, as much as the *SS Torrey Canyon* disaster of nine years earlier and the *Exxon Valdez* of four years later, worked as a catalyst for analyses on the role of scientific experts. This time, however, rather than naively arguing that environmental disasters highlighted a growing demand for more experts to tackle specific issues, new literature argued for a sociological understanding of risk and risk management. Recognizing that experts operate in highly uncertain domains, this literature called for larger investments on understanding the social processes empowering them rather than taking for granted that advanced societies need more experts to tackle environmental degradation.[72] Needless to say, these sociological analyses occupy little space in the CCMS program.

By 1991, the committee had completed thirty-six studies and thirty-three were then underway. Some of them had undoubtedly importance for addressing social issues arising in areas as diverse as health-care provision, (nonnuclear) waste treatment, earthquakes and disaster assistance, and even archaeological recuperation on top of those directly concerning pollution that previous chapters have examined in greater depth.[73] Yet the end of the Cold War focused the committee on actions in fields that they had deliberately overlooked in previous years, such as the environmental impacts of military installations. This is remarkable, especially because the committee had deliberately ignored it when the scholar Patrick Kyba openly criticized the CCMS in 1973.[74]

The committee's new agenda can be explained once again in terms of political expediency. First, the new research focus gave NATO officials an opportunity to deflect public criticism. Moreover, in the early 1990s, the alliance's officials had had to deal with the distressingly large number of abandoned military installations in Eastern Europe in the aftermath of regime change. These installations presented a variety of unnerving security issues, being managed by poorly paid military personnel, exposed to looters and intruders, and containing a variety of pieces of military equipment and weaponry for which no inventory in the West existed. By then, furthermore, a substantial reduction of nuclear weapons was finally within sight as result of the 1987 Intermediate-Range Nuclear Forces treaty granting a substantial withdrawal of warheads in the Soviet Union and Western Europe. Although a number of Western intelligence agencies dealt directly with the problem of establishing (or restoring) safer conditions for the management of these military bases and missiles, one way to facilitate access to these installations for NATO personnel was to make former enemy governments wary of their environmental impacts.

In 1987, the CCMS launched a pilot study on this topic, suddenly recognizing, fourteen years after Kyba's denunciation, that environmental conservation and sustainability could be seriously undermined by NATO's military wastes. Through the German-led pilot study "Promotion of Environmental Awareness in the Armed Forces," the CCMS instigated a self-reflective effort to look into ways to mitigate their environmental impacts. In 1990, the CCMS pilot project, coupled with another one on defense environmental expectations, led NATO to elaborate environmental policies for its commands and units.[75]

Yet, by the time these initiatives were taken, the change of regime in Eastern European countries was in full swing. This climate helped the CCMS provide the means and rhetorical justification to take the demobilization process then occurring in former Soviet satellite countries under its wing. In 1992, Germany and Norway jointly launched the pilot "Cross-Border Environmental Problems Emanating from Defence-Related Installations and Activities," seeking to promote cooperation in surveying, assessing, and preventing contamination in critical areas, such as the Barents Sea, the Baltic Sea, and the Black Sea. Two years later, the US and (now reunited) German delegation launched the "Environmental Aspects of Reusing Former Military Lands" pilot project with other eighteen NATO nations. The project entailed exploring the reuse or restoration of former bases in Eastern Europe such as the Ralsko training ground in Czech Republic, the Amari air base in Estonia, and the Liepaja naval base in Latvia. This flourishing of activities culminated in the organization of the workshop "Military Activities and the Environment" in Warsaw (Poland) to promote a dialogue between old NATO partners and new nations in Eastern Europe seeking to gain access within the defense alliance to the best methods and approaches to military-related environmental issues.[76]

The CCMS initiatives helped NATO to set a foot in former enemy countries (and learn more about their decaying military systems), while also creating favorable circumstances for political integration. The intelligence and diplomacy returns of these missions paralleled the revision of NATO's strategy. Its new strategic concept recognized that security required a broader analysis of a variety of factors, including environmental ones.[77]

It is of note that while in previous occasions (i.e., the 1972 Stockholm Conference) the CCMS had rejected opportunities for recognizing other international organizations, in 1992 it finally endorsed the 1987 Montreal Protocol on the phasing out of ozone-depleting substances on the occasion of the protocol's forthcoming Conference of Parties in Denmark. Recognizing that every NATO country was a party to the protocol, the CCMS message stressed that "NATO supports the accelerated phase out of these substances."[78]

That said, NATO's intervention in the Balkan conflict presents a startling contrast to the imperative of environmental awareness. NATO thus continued to be inconsistent in practicing and preaching environmentalism. One bone of contention about the alliance's intervention in the conflict was the use of depleted uranium shells. Although these uranium projectiles have the merit of penetrating ever thicker layers of armor because of the density of the shell material, their environmental and health impact is notable in that they increase the amount of radioactivity in the areas where they are used. The radioactive contamination of many areas of the former Yugoslavia was thus the most problematic legacy of NATO's intervention in the Balkans. This result was all the more paradoxical in light of the fact that when the bombing adding to the radioactivity of Balkan soil was agreed upon, the Science Committee was about to sponsor research on the environmental impacts of increased ultraviolet radiation deriving from the thinning of the ozone layer.[79] Depleted uranium shells also were tested at the polygon of Salto di Quirra—once the main Italian facility for new types of missiles and now a dump for military waste where no provisions for environment and public health existed.[80] Crucially, a NATO environmental impact study of depleted uranium used in the Balkans was carried out only after the public scandal that their use produced.

But this was not the only case to emerge in which key environmental and health principles had found little space in NATO's military protocols. In 1998, the alliance shut down one of the former NADGE installations in Italy situated in an underground bunker under Mount Venda, in the Veneto region. And, since then, a number of inquiries have revealed that high concentration of radon and asbestos have been responsible for the premature death of forty of NATO's former military staff. One health official of the Italian air force and the Italian Ministry of Defense have been recently condemned for how health and safety were disregarded in the daily running of the Italian base.[81]

These contradictions recall that the CCMS promotion of environmental actions was aligned with NATO's use of the environment as a diplomatic device. This meant that when environmental initiatives conflicted with the alliance's political and defense agendas, they were at times discarded or overlooked, and when they instead could bring additional diplomatic and intelligence gains, they were enthusiastically endorsed. In the 1980s, NATO's intervention in environmental affairs proved to be more than ever grounded on "establishment" politics, the search for solutions to environmental issues with the diplomatic ambition of maintaining the status quo while reducing the influence that more radical groups and organizations could play in environmental affairs.

This was the case for NATO's science program too. The Science Committee opted for an investment that overtly reaffirmed the political uses of science, through the Science for Stability program, and rejected calls for a major investment in climate change research, notwithstanding the promotion, for the previous ten years, of environmental studies that heralded NATO's intervention in the understanding of natural and ecological processes and the human impact on nature. This chapter has explained this contradiction by examining contextual political factors, including the implications of the climate change discourse (i.e., through the nuclear winter scenario) in broadening the criticism of NATO as a defense alliance in a critical moment of the Cold War. Of course, other important issues have to be taken into account, including Chabbal's personal agenda in fostering earth science studies. But NATO's winding road toward environmentalism and environmental studies is now plain to see.

In 1985, the scientist who could be considered the program's undisputed leader and most significant contributor, Isidor I. Rabi, finally left the Science Committee. Those who replaced him struggled to keep the program going. NATO's sponsorship of science was still very alive at the beginning of the twenty-first century, but it went through a transformative process, as the epilogue to this volume will show.

EIGHT

Epilogue: An Evergreen Alliance?

In the last twenty-five years, NATO's international role has changed dramatically. The alliance's administrators could not foresee the crumbling of the Soviet Union, nor had they thought about NATO's mission once the bipolarity that characterized the Cold War was upset forever. The dissolution of the Soviet bloc made many, within and outside the alliance, wonder whether an organization like NATO explicitly set up to counter the threat represented by the Soviet Empire still had reason to exist.[1]

While facing new challenges, not only did NATO reaffirm its role as a leading multilateral defense organization, but expeditiously moved toward its enlargement, welcoming new members and endorsing a new strategic concept. Its transformation into a new entity with new imperatives, which is still happening, makes some of the questions that this book has dealt with all the more important. Is there someone still interested in "greening" the alliance sixty years after the Science Committee's establishment and fifty after the CCMS's birth?

This concluding chapter cautiously responds to this question while recalling some of the book's overarching themes. It shows that the pursuit of enlargement has, until very recently, represented the new compass for steering the alliance's science and environmental programs, also shaping its sponsorship activities. In particular, the building of relations with new members and potential allies worldwide has defined new patronage strategies. But an appraisal of

NATO's current posture may coincide with the end of the alliance's traditional sponsorship agenda. I conclude that new environments, such as the complex of electronic networks typically identified as the "cybersphere," represent today a more significant research focus than what the real environment has been in the past. Even if one might speculate that NATO's environmental phase has approached its twilight, this epilogue underscores the importance of reflecting on the half-century history of the alliance's greening to better understand how it informs current approaches, research practices, and policies.

No Longer Greening? NATO's New Patronage Strategy

NATO's longevity as a defense alliance appears in stark contrast with that of similar organizations established during the Cold War. Long gone are the days of its twin organizations on other continents: the South-East Asian Treaty Organization (SEATO) was disbanded in 1977, and the Central Treaty Organization (CENTO) dissolved two years later. In 1991, the Warsaw Pact folded too, and, in 2011, so did the Western European Union (WEU; despite rumors in the 1990s about its resurgence). By contrast, not only did NATO survive the end of the Cold War, but the termination of competing multilateral defense projects (especially the Warsaw Pact and the WEU) further invigorated it. NATO also has acquired greater influence internationally by offering membership to former Soviet satellites in Eastern Europe and contributing to a number of international coalitions.

At the end of the century, very few officials and analysts had a clear understanding of whether the alliance should open membership to Eastern European states. In 1997, the US ideologue behind the doctrine of containment, George Frost Kennan, questioned the strategic vision behind its enlargement, judging it a "fateful error." It would have unnerved the Russians and produced a second Cold War, he claimed.[2] Yet the Balkan conflict, and the role that NATO played in its unfolding, helped its administrators to affirm the alliance's pivotal role above and beyond traditional Cold War affairs. By 1999, when ex-Yugoslavia's territory was bombed by NATO air forces, the administrations of the Czech Republic, Hungary, and Poland had agreed to join the alliance. A few years later, even the strategy of eastward expansion was to be superseded. Although nine more Eastern European countries joined NATO (Albania and Croatia the most recent cases), the events that followed the 9/11 (2001) terrorist attack on New York's twin towers propelled NATO into the global arena. Terrorist actions by Islamic extremist groups paved the way for the alliance's missions in Afghanistan (2003), Iraq (2004), and Libya (2011). The rise of religious terrorism eased NATO's transition from an organization tied to Cold War Europe to one that could play a new role in conflict resolution globally.

"NATO was able to successfully reinvent itself, to change with the times before obsolescence risked to take over" claims the analyst Jamie Shea.[3] The establishment of international alliances using the device of regional cooperation also revealed NATO's quest for global reach. These coalitions include today the Partnership for Peace agreement (with former Soviet republics); the Mediterranean Dialogue (with North African partners), and especially the Euro-Atlantic Partnership Council, which currently comprises fifty countries, including the twenty-eight NATO member states. NATO is no longer the one of two Cold War defense alliances that survived the conflict's end, but, arguably, one of the most powerful multilateral defense organizations on the planet.

NATO's effort to adjust to the rapidly changing geopolitical landscape of the post–Cold War period also resulted in the search for a new strategic concept. If McNamara's flexible response was the alliance's key doctrine for twenty-three years, then its successor doctrines lasted for a much shorter period of time, confirming NATO's state of flux. In 1991, the Western alliance launched a new doctrine to strengthen dialogue and cooperation with former Cold War competitors. It was replaced, eight years later, by one underlining the need for NATO to enlarge. The straining of relations with Russia that Kennan had foreseen set the stage for one more change of direction. In 2010, following the initiative of former NATO Secretary General Anders Fogh Rasmussen, NATO adopted "Active Engagement, Modern Defence," which finally recognized the alliance's role as a global player. The present strategic concept marks the end of previous efforts to integrate the Russian Federation into the alliance, but, more importantly, no longer identifies geographical areas of influence and suggests that NATO's forces could now reach everywhere on the planet to counter perceived threats.[4]

While in these years of transition, NATO officials attempted to align the alliance's patronage of science and technology to its changing strategy and imperatives, it would be impossible to document in detail the projects that NATO has recently funded. But it is important to note that the alliance's science administrators have continued up until very recently to use patronage as a diplomatic device. Echoing the original conviction of the Three Wise Men that cooperation in science and technology strengthens NATO relations, and giving furtherance to initiatives identifying the promotion of science as a stabilizing force in international relations, they have posited that sponsorship of scientific and environmental programs could produce new synergies. They have now, however, further bolstered science diplomatic efforts to befriend potential allies in distant places that are the focus of regional conflicts. In recent years, NATO has sought to boost global scientific and technological capability, especially in countries where it had found resistance in the knowledge that sponsorship could grant political returns. Opportunities to exercise science diplomacy materialize especially when the scientists that the alliance endorses end up in government (or play a role in their government's advisory committees) and contribute to shaping national policies. NATO's seed money thus still has borne diplomatic fruits, as it were.

This approach reiterates what *Greening the Alliance* has shown to be one of the pillars of the long-lasting science program established at NATO: namely, the funding of science projects in a process of constructive evasion. In the last sixty years, NATO sponsorship has aimed to smooth international relations by removing diplomatic tensions between countries. In NATO's history, this strategy was especially effective when tensions between allies materialized, such as those on NATO's trajectory as a multilateral power organization with an uneven distribution of power among members; the disagreements on the alliance's strategic posture (including its nuclear and nonnuclear deterrents); and the temporary renunciation of key military responsibilities by leading member states (e.g., France). In this respect, this book has shown NATO to be an organization whose existence has constantly been undermined by tensions. But it also emphasizes its role as a sponsor for science and environmental initiatives worked as a countervailing force; the administration of scientific affairs has functioned, essentially, as a form of parallel diplomacy.[5]

The tacit diplomacy game embedded in NATO scientific affairs was decisive also in forging new coalitions between its member states and giving new directions to the alliance throughout its history. That between US and UK delegations was decisive in setting up a science program, and the hidden synergies between French and British officials nearly killed it five years later. The alliance between less developed NATO countries signaled the emergence of new tensions with regard to power imbalance within the alliance and the unity of intents between the United States and West Germany forged the science program's turn toward environmental research. Although there is no doubt that the United States continued to have a leading role throughout NATO's history, this book tempers views on American hegemony in revealing the importance of these fleeting alliances in shaping it. At times, other variables further complicated the diplomatic game even further, such as the personal relationships between representatives, the differences between similarly named government departments, the interests of the private sector, and the competing roles of other international organizations.

That said, the end of the Cold War marked a substantial change in NATO's exercise of science diplomacy. In 1997, an independent high-level group comprising scientists from Portugal, Norway, Italy, France, Germany, the United Kingdom, and the United States concluded that NATO's science program should no longer have an "intra-alliance" focus but concentrate on "alliance-partner" collaborations.[6] Since the NAC endorsed these recommendations, at the turn of the century the now renamed Security Through Science program was divested from its original ambition of further integrating the alliance and became instead a device to propel friendly relationships with partner countries and potential partners around the world. These included, first and foremost, countries in Eastern Europe and the Mediterranean, in line with NATO's enlargement strategy. But it targeted especially the newly formed Russian Federation, which at this point in time NATO wished to establish closer relations with through the creation, in 2002, of the NATO-Russia Council, and, from April 2003, the joint research activities with Russian scientists carried out under the aegis of the council's Science for Peace and Security Committee. This collaborative work eventually ended up taking a considerable share of NATO's investment in science and technology and even entailed the pooling of knowledge on sensitive research items such as, for example, defense against chemical, biological, and radiological agents.[7]

In 2005, the diplomatic uses of science even convinced NATO's administrators to reconsider their traditional disinclination to collaborate with other international agencies, as shown by the Environment and Security Initiative (ENVSEC), which began in 2003 with the goal of producing environmental security assessments in three world regions (Central Asia, the southern Caucasus, and southeastern Europe) hit by the collapse of the Soviet Empire and made even more vulnerable by the resurgence of nationalist and religious tensions. The program focused on improving the use of natural resources and stabilizing the targeted regions through successful environmental management strategies. Originally established by the UNEP, the UN Development Program, the Economic Commission for Europe, and the Organization for Security and Co-operation in Europe, ENVSEC welcomed NATO among its sponsors. In turn, this led to new studies covering subjects such as the analysis of seismic risk in the Caucasus region, the study of Caspian ecosystems, and a survey of chemical and radioactive contamination in proximity of defense sites such as the nuclear polygon of Semipalatinsk.[8]

Although the interaction between science and environmental projects had typified NATO's science program for more than twenty years, it now consolidated through the creation, in 2006, of the Science for Peace and Security (SPS) program, merging the Science Committee and the CCMS and administered through NATO's Public Diplomacy Division (the Scientific and Environmental Affairs Division was disbanded). The new program, as the name suggests, focused on research and actions in conflict prevention. In particular, because of the growing availability globally of chemical, nuclear, and biological agents, the SPS program stimulated the study of weapons' detection and protection strategies.[9] In 2009, the NATO agency also set up the Defence and Environment Expert Group to further analyze the environmental impacts of military activities in continuity with previous CCMS efforts.[10]

The SPS advanced the alliance's promotion of research projects in distant places in order to increase its influence outside Europe, as demonstrated by the completion, in 2009, of "SILK—Afghanistan," a project set up to provide high-speed Internet connection via satellite and fiber-optic underground cables to eighteen Afghan universities and to governmental offices in Kabul in the aftermath of the conflict with the Taliban. It was completed thanks to the availability of a preexisting network (the Virtual Silk Highway) that in previous years NATO had set up in the South Caucasus.[11]

But the alliance's state of flux, and the rapidly changing geopolitical landscape in which it operated, have recently produced more changes. In 2010, after the adoption of the new strategic concept "Active Engagement, Modern Defence," the SPS program was transferred again, this time to a newly established division called Emerging Security Challenges.[12] The renovation paralleled a deep reorganization of NATO's defense research. In 1997, the DRG and AGARD had already merged into a new division, which was also disbanded in 2010 and replaced by the NATO Science and Technology Organization (STO). It comprises, notably, the former marine research center SACLANTCEN (renamed as Centre for Maritime Research and Experimentation).

The creation of the new Emerging Security Challenges division marked a decisive reorientation of the alliance's "third dimension" and the definition of a new research agenda closer to the alliance's strategic priorities. Although this reorganization did not entail a renunciation of a mixed philosophy (e.g., using science patronage as a device for both diplomacy and improving defense),[13] it means that current support is now geared more toward defense-oriented research, with the SPS program being "one of the principal stakeholders in NATO S&T."[14] In 2011, recognizing that "our strategic environment has probably never been as uncertain," NATO's strategic analyst Massimo Panizzi highlighted the importance of countering this unpredictability, globally, with new research on terrorism, cyberattacks, and the proliferation of weapons of mass destruction.[15] The new division's officials have also been looking, with growing attention, to the volatile issue of energy supply in light of the security challenges that its uneven distribution may lead to. Six subsections and a directorate now actively deal with these new challenges, and the SPS is integrated in these activities.

This reorganization means that sponsorship of traditional scientific research and environmental actions may no longer be a priority in the alliance's future. First, the stationary level of allocations for research indicates a wish to target interventions on specific research areas rather than launching new major programs. In 2013, the SPS sponsored research projects whose costs totaled 12.1 million euros. This budget represents half of the 2000 SPS budget (24.3 million euros) and is hardly comparable to that of other multilateral research agencies, such as the European Science Foundation (52 million) or the European Research Council (1.5 billion).[16] In fact, it hardly comes close to that of some national research organizations (the UK Natural Environmental Research Council's budget is higher).[17]

Moreover, the significant reduction in the number of SPS applications suggests that the alliance's grants program may have lost some appeal. Applications have more than halved from an average of approximately five hundred per year before 2010 to one to two hundred in the period 2011–14. This is remarkable especially because the overall budget available for the 2014 program has decreased only marginally (from 13 to 12 million euros). Yet the number of awards has plunged, too (from 127 in 2008 to 51 in 2013 and 74 in 2014). Despite growing funding opportunities, fewer applicants have succeeded in being awarded a NATO research grant [18]

Even more problematic is the decreasing number of environmental security activities sponsored by the alliance. The relevant data (42 in 2008, 39 in 2009, 24 in 2010, 3 in 2011, 7 in 2012, 1 in 2013, 3 in 2014) display a negative trend. And, in contrast with this trend, the demand for support of environmental actions is currently growing, especially from partners in the Euro-Atlantic Partnership Council and the Mediterranean Dialogue.[19]

One important reason for this decrease is the deterioration of NATO's relations with the Russians following Russia's annexation of Crimea and the conflict with Ukraine. On April 1, 2014, a statement of NATO's foreign ministers announced the suspension "of all practical civilian and military cooperation between NATO and Russia" which in turn entailed the end of collaboration on science and environmental projects.[20] It is notable that since then the bulk of the SPS program has been reoriented toward Ukraine and research areas of interest to its administration including telemedicine, demining, and remediation of polluted military sites.[21]

To sum up, the data available suggest a reorientation of NATO's efforts, and, presumably, a somewhat reduced interest in furthering scientific and environmental research. Recent SPS reports state that the organization wishes to strengthen scientific collaborations (especially with Ukraine, Moldova, and Georgia) and partners in the Middle East such as Jordan and Iraq.[22] The current research priorities and budget will very likely allow carrying forward these plans. But gone are the days when major NATO oceanographic surveys made other researchers in Western Europe anxious because the alliance could lavish funds for a selected number of specialists, also upsetting scientific relations across the Iron Curtain. It is thus important to consider what fifty years of sustained support to environmental studies and actions have meant in terms of legacy. What difference has this investment made? What has changed, and what hasn't, since the times of NATO's primacy in the promotion of science and environmental actions?

NATO's Programs and Their Legacy

Greening the Alliance has shown NATO to be among a handful of organizations pioneering environmental consultations like the meeting in Durban that opened this volume and the Twenty-First UNFCCC Conference of Parties in Paris, giving leeway to, allegedly, the "first universal climate agreement."[23] Today's talks share much with the early CCMS negotiations insofar as they are typified by the search for urgent solutions to a compelling issue requiring complex adjustments in living standards, production and storage of energy, and much more.

In contrast, however, NATO's early approaches to environmental policymaking were typified by a distinctive self-assurance in Western-centric, science-based methods to deal with environmental issues. And underdeveloped countries in Europe and elsewhere were even considered too slow in catching up and putting science and technology at the service of environmental policymaking in the way advanced Western nations had done. Facing the oil slicks of the *Torrey Canyon* disaster, NATO officials showed the same buoyancy that Doctor Teller displayed when discussing the effects of nuclear fallout, a self-assurance paralleled by Russell Train's confidence in the ability of humankind to successfully address air pollution problems within the decade. This book has documented the journalistic stunt that in 1976 saw the US environmentalist and NATO Secretary General Joseph Luns promote the clean-engine automobile while keeping its design's hiccups hidden.[24] Fifty years later, we know that the introduction of electric vehicles on the global market still faces imposing difficulties.

One would also hope that future environmental agreements will not be undermined so much by policymaking on environmental protection in the presence of compelling national imperatives and vested interests. If Nixon's presidency heralded the promotion of international environmental conventions (mainly thanks to Train's intervention), his strategy consisted of expounding the virtues of environmental protection in an effort to direct the alliance toward a unifying agenda at a critical junction in NATO's history. And Nixon's allies manifested, in open and restricted talks, decisive reservations. Environmental issues that NATO wished to tackle were thus deliberately evaded, even though the alliance's media releases portrayed it as pioneering environmental protection. This book has shown instead that allies hesitated over, or even rejected, solutions that could have been effective. Though openly publicizing its new "green" agenda, for more than a decade they overlooked, for instance, the environmental impacts of the alliance's military operations and exercises.

Moving from past to present, NATO's position with respect to forthcoming environmental negotiations is as hazily defined as its current investment in environmental actions. Twenty-eight NATO allies are all UNFCCC signatories and are responsible for 32.32 percent of global carbon emissions. This makes the alliance a key stakeholder.[25] Despite this, NATO's officials have been incredibly shy on the Paris agreement. On October 11, 2015, just before UNFCCC delegates convened, the NATO Parliamentary Assembly stressed the importance of reaching an agreement and recalled its merits especially as a matter of international security.[26] On that occasion, the NATO Secretary General, Jens Stoltenberg, avowedly supported the assembly's resolution. Yet there has been no follow-up to these statements, not even after the Paris agreement was signed. As to endorsing the Parliamentary Assembly's resolution, NATO officials have been equally noncommittal, even though the resolution explicitly called for increasing the "frequency of military and political consultations on climate change within NATO, including at NATO summits."[27]

Environmental research also has dramatically changed in recent years. We have seen that in the 1990s former NATO Science Committee members played a very distinctive role in the climate change debate, explicitly arguing against recent claims on global warming. This is not just proven by the quizzical initiatives of French science administrators

Robert Chabbal and Claude Allègre, but also by what Oreskes and Conway have identified as Nierenberg's and others' "campaigns to mislead the public and deny well-established scientific knowledge [on climate change] over four decades"[28]

Looked at from the perspective of this book's narrative, the polarization defined by the climate change debate that Oreskes and Conway so cleverly describe is more than just a clash with doubtmongers—a group notably including the former chair of the Science Committee, William Nierenberg. The pages of this book suggest their skepticism to be deeply rooted in their stances as former Cold War scientists and prominent figures in NATO's Science Committee.

Nierenberg shared with NATO peers an understanding that scientists could successfully address environmental problems as they had successfully tackled critical Cold War tasks. As the highest strata of the atmosphere, the deepest sections of the oceans, and anything in between were probed through sophisticated electromagnetic equipment and sensors, the science they propounded provided assurance about the possibility of more accurately detecting incoming air and naval threats, whether they were submarines, aircraft, or missiles. This confidence on the potential of NATO's detection networks engendered optimism in the alliance's ability to manage open conflict between superpowers. And it ran counter the bleak projections of Solly Zuckerman about a nuclear holocaust in Europe.[29]

This enthusiasm has been vividly recalled in the volume through the fantastic propositions of Edward Teller. Although his schemes were never endorsed by Rabi and Zuckerman (and presumably rejected by Nierenberg too), they displayed ingenuous expectations about humankind's ability to put environmental change under control. Nierenberg's generation of techno-enthusiasts rejected bleak forecasts about nuclear holocausts and nuclear winters, recognizing that more accurate forecasts would pave the way to convincing policies. In the 1960s, it was the buoy-installed current meter that found application in both reconnaissance-related studies and fishery research; in the late 1980s, it was the sophisticated satellite-mounted Synthetic Aperture Radar that could assist both navy officials controlling nuclear submarines and environmental administrators eager to monitor oil spills.[30] The shaping of these novel scientific exercises as defense operations reassured their architects about man's quest to know the environment, to improve defense in general through this knowledge, and to put environmental phenomena under human control.

So it is no surprise that NATO's entrance into the realm of environmental analysis, management, and policymaking was also marked by this distinctive confidence, as shown by Rabi's speech on the occasion of the tenth anniversary of the Science Committee, in which he hinted at the prospect that scientific analyses of environmental degradation be informed by the experience, practices, and mindsets of Cold War scientists. NATO's investment also was premised in the attempt to overwhelm grassroots environmental movements busy denouncing the limits of modern science and technology in addressing environmental issues, which meant that the environmental anxieties produced by the civilian applications of nuclear energy were never debated enough at NATO. It is worth noting that it was yet again the climate skeptic Nierenberg who prevented it while labeling antinuclear protesters as "a vociferous minority."[31]

Of course, Nierenberg was not alone in propagandizing the notion that environmental degradation could be fixed. Other NATO science administrators such as Eduard Pestel, Gunnar Randers, and Nimet Özdas helped to configure solutions to environmental problems as novel tasks for NATO-sponsored environmental scientists. Patronage engendered optimism in environmental and surveillance matters alike. An increase in surveillance appeared to offer a way out of nuclear Armageddon, insofar as prompt detection of enemy threats could instigate accurate responses. Improvement in environmental monitoring granted solutions to environmental crises too. If today we place emphasis on the issue of environmental monitoring when looking for answers to global environmental challenges from the available literature, it is partly because we draw on research practices that NATO's patronage contributed to in a number of laboratories, especially in Western Europe.

Although this has happened, much has changed in the landscape of environmental research. The distinctive confidence shown by the pioneers of NATO's environmental work now appears somewhat less justified; while finding solutions to specific issues, scientists have uncovered a number of problems invading the political domain. There is a flourishing sociological literature today discussing what prevents a successful dialogue between scientists on global environmental challenges. Silvio Funtowicz and Jerry Ravetz have portrayed today's science as postnormal: "facts are uncertain, values in dispute, stakes high and decisions urgent."[32] Mike Hulme has shown, against the backdrop of Funtowicz and Ravetz's analysis, that there a number of entirely legitimate reasons today why environmental scientists fail to produce consensus on key environmental issues. Contrary to what many at NATO believed, improvements in monitoring and increases in accuracy in the use of remote-sensing equipment has returned the conviction that scientific knowledge "will always be incomplete, and it will always be uncertain."[33]

Changes in the patronage of environmental studies may have contributed to these transitions. At NATO, relatively small groups of researchers busy studying environmental phenomena could tap fairly steady (and at times steadily increasing) funding streams. These funds reassured them about their research findings and provided them with the distinctive self-assurance that this book has portrayed. But, in the last fifty years, the funding landscape for the environmental sciences has changed. The end of the Cold War has eroded the funding of traditional environmental studies as configured through the remote sensing of specific physical parameters. NATO's much-reduced presence on the sponsorship scene today is part and parcel of a broader reconfiguration, as a variety of sponsors, nationally and internationally, offer support for novel studies but recognize the merits of interdisciplinary work and the complexity of natural environments. Environmental monitoring exercises of the last thirty years have returned a wealth of data on the planet and its features, but rather than nourish a naive optimism on the ability to control nature, they appear to have reduced the confidence that we can aspire to know it in full.

The impasse of environmental scientists who investigated the *Cavtat* incident demonstrates that NATO scientific missions eventually made environmental scientists more alert to this transition. The interaction between various experts from physical oceanography, chemistry, and biology led to their realization that while sunken lead compounds could be harmless, their entering the food chain could cause harm. But they also understood that the amount of harm that would be caused was impossible to predict. Although they took responsibility for reassuring the populace about risks, they realized the outcome of their exercise was highly uncertain.[34]

So NATO's sponsorship of environmental studies might have contributed to defining these funding patterns and the pioneering of novel, collaborative approaches to research in Western Europe. But today's increase in the monitoring of environmental phenomena no longer provides reassurance of the possibility to successfully address crises. Nor has it, actually, provided confidence about surveillance matters, either.

A New Environment for the Surveillance Imperative

The words of Massimo Panizzi recalled earlier in this epilogue show that *uncertainty* is no longer in the vocabulary of environmental scientists alone but has been adopted by NATO strategic analysts too. His lack of confidence in the ability of the alliance to promptly recognize, detect, and counter incoming threats may seem puzzling in the face of the massive investment in surveillance equipment and research that has typified the alliance throughout its history.

So where does uncertainty come from? In the late 1970s, when NATO's greening was in full swing, US scholar Robert Jervis was busy developing the notion of the "security dilemma." Jervis explained that "as each state seeks to be able to protect itself, it is likely to gain the ability to menace others."[35] Although he developed the concept mainly as a reflection on the escalation typifying the nuclear arms race, over time the notion has found broader application, also becoming one of the conceptual pillars of surveillance studies. Today we explain the evolution of modern means of surveillance in light of the security game of increasing protection and menace that Jervis astutely described.

NATO's history is rich in examples proving Jervis correct: through the surveillance of nuclear submarines and supersonic aircraft and missiles, the Cold War conflict unfolded as a game defined by the succession of protection from a new menace that novel surveillance equipment provided, and the menace that it represented to the enemy's means of protection. Chapter 6, for instance, showed remarkable changes in NATO's defense research sponsorship as resulting from efforts to reconfigure the surveillance networks of the alliance, especially through the setting up of NADGE.[36]

The current reorientation of NATO funding reveals yet another important stage in the evolution of the security dilemma. No other defense alliance in the world today could successfully threaten NATO. The Salala incident discussed at the beginning of this book may display the intelligence shortcomings that have typified a recent NATO mission, but it also illustrates the alliance's surveillance and counterattack capabilities. The possibility exists today for its military authorities to manage conflict remotely through weapon systems employing drones on the other side of the planet and successfully directing them toward targets, with virtually no risk to the alliance's forces.

Certainly the natural environments still harbor many secrets, but these may well no longer be the chief hindrance to NATO operations, which may explain why NATO's investment in environmental research is changing (possibly shrinking). A recent article recalls the circumstances of the alliance's defense research structures and the current decline of the NATO marine research center in La Spezia. Twenty years ago, nuclear submarines could stay hidden underwater for months and represented a noticeable (yet invisible) threat to the alliance. Today, underwater drones make it impossible for even the most sophisticated nuclear submarine to evade surveillance. NATO has invested less in marine research, but sea drones have called for larger investments.[37] And it is still today's drones, rather than undercover oceanographic vessels such as the *USS Pueblo*, that are responsible for sea reconnaissance operations.[38]

The alliance is now investing much more on cyberdefense. NATO's rapidly growing funding of cybersphere research suggests that the natural environments no longer represent the unknown frontier. Systems devoted to the surveillance of electronic spaces have expanded. NATO's electronic environment was conceived in support of its surveillance network and was accessible only by a relatively small number of accredited operators. The evolution of computing and web-based systems has upset this original design and produced larger networks that are more vulnerable. The new cyber environment is the chief focus of NATO's security and has occupied a prominent space in its patronage too. The anxiety of its military commanders derives primarily from the realization that the alliance's weapon systems could be undermined by tampering with their electronic infrastructure. In October 2014, several newspapers reported that hackers allegedly paid by Russia had successfully managed to infiltrate NATO's email system and gained access to restricted information.[39] More recently, these accusations have hit the US administration, with federal intelligence agencies claiming Russian interference in the US presidential election campaign.[40]

It is presumably because of these new threats that NATO's science program has changed. We have seen that the task of research on cyberdefense is one of the main priorities of the new Emerging Security Challenges division. Its defense research equivalent, the STO, has also invested heavily in cyberspace, and in 2010 a new Cooperative Cyber Defence Centre of Excellence was established. Based in Tallinn (Estonia), the center hosts a small team that looks into cyberattacks that could undermine NATO's defense systems.[41]

New exercises, like Locked Shields, differ dramatically from their predecessors, including—for instance—*Able Archer-83*. They have the ambition of understanding how NATO could continue using the alliance's defense network while facing a cyberattack. There is thus a clear perception that even its defense role is bound to change dramatically, as preventing an enemy's attack is no longer a matter of military preparedness and up-to-date information about environmental conditions, but of how to manage the electronic data infrastructures that are the pillars of NATO's modern defense equipment.[42] In the recent Locked Shields, NATO computer experts have been divided into two groups: one—a Blue Team defending the imaginary Berylia—was called in to repel the attacks of the Red Team's Crimsonia. The exercise lasted for two days in order to verify that the alliance could carry out its traditional shield functions while being under attack.

The cybersphere is witness to a variety of environmental threats, but these take the form of computer bugs and lines of code, rather than storms and waves. The alliance's investment in cyberdefense is escalating, and so has been that of NATO allies. So the novel nature of surveillance defines NATO's future defense setup, and the colors used in Locked Shields take us back to traditional Cold War exercises when blue forces confronted the red menace. But one color that has been distinctive of its fifty-year-long scientific patronage now appears to be missing. Is the alliance no longer greening?

Notes

INTRODUCTION

1. Dag T. Gjessing, *Remote Surveillance by Electromagnetic Waves for Air-Water-Land*, Ann Arbor, MI: Ann Arbor Science, 1978.
2. See, for instance, Jan van der Bliek, ed., *AGARD: The History, 1952-1997*, Illford, UK: RTO/STS Communications, 1999; Donald Ross, "Twenty Years of Research at the SACLANT ASW Research Centre, SACLANTCEN," Special Report M-93, January 1, 1980, La Spezia, Italy: NATO, 1980; Morris Honick and Edd M. Carter, *SHAPE: The New Approach, July 1953–November 1956*, Brussels, Belgium: SHAPE, 1976. An interesting book with a comprehensive chapter on NATO science is Francis A. Beer, *Integration and Disintegration in NATO: Processes of Alliance Cohesion and Prospects for Atlantic Community*, Columbus: Ohio State University Press, 1969, 204–39.
3. John Krige, *American Hegemony and the Postwar Reconstruction of Science in Europe*, Cambridge, MA: MIT Press, 2006. Some of these concepts have been recently recalled in Krige, *Sharing Knowledge, Shaping Europe*, Cambridge, MA: MIT Press, 2016.
4. Jacob Darwin Hamblin, *Oceanographers and the Cold War: Disciples of Marine Science*, Seattle: University of Washington Press, 2005; Hamblin, *Arming Mother Nature: The Birth of Catastrophic Environmentalism*, Oxford: Oxford University Press, 2013; Hamblin, "Environmentalism for the Atlantic Alliance: NATO's Experiment with the 'Challenges of Modern Society,'" *Environmental History* 15 (2010): 54–75.
5. A wider range of examples from other disciplines are presented in Simone Turchetti and Peder Roberts, eds., *The Surveillance Imperative: Geosciences during the Cold War and Beyond*, New York: Palgrave, 2014. For cartographic and geodetic research, see William Rankin, *After the Map: Cartography, Navi-*

gation, and the Transformation of Territory in the Twentieth Century, Chicago: University of Chicago Press, 2016.

6. S. M. Amadae, *Rationalizing Capitalist Democracy: The Cold War Origins of Rational Choice Liberalism*, Chicago: University of Chicago Press, 2003; Paul Erickson, Judy L. Klein, Lorraine Daston, Rebecca M. Lemov, and Thomas Sturm, *How Reason Almost Lost Its Mind: The Strange Career of Cold War Rationality*, Chicago: University of Chicago Press, 2013; Jamie Cohen-Cole, *The Open Mind: Cold War Politics and the Sciences of Human Nature*, Chicago: University of Chicago Press, 2014. On NATO operations research, see also chapter 8 of Krige, *American Hegemony*, 227–52. On its application to global problems, see Thomas Robertson, *The Malthusian Moment: Global Population Growth and the Birth of American Environmentalism*, New Brunswick, NJ: Rutgers University Press, 2012.

7. The ancestry of geophysical research is discussed in Ronald Doel and Naomi Oreskes, "The Physics and Chemistry of the Earth," in *The Cambridge History of Science*, vol. 5, edited by Mary Jo Nye, Cambridge, UK: Cambridge University Press, 2012, 538–57. See also Charles C. Bates, Thomas Frohock Gaskell, and Robert B. Rice, *Geophysics in the Affairs of Man: A Personalized History of Exploration Geophysics and Its Allied Sciences of Seismology and Oceanography*, New York: Pergamon, 1982.

8. Paul Forman, "Behind Quantum Electronics: National Security as Basis for Physical Research in the United States, 1940–1960," *Historical Studies in the Physical and Biological Sciences* 18 (1987): 149–229.

9. Daniel Kevles, "Cold War and Hot Physics: Science, Security and the American State, 1945–56," *Historical Studies in the Physical and Biological Sciences* 20, no. 2 (1990), 239–64. On the debate, see Naomi Oreskes, "Science in the Origins of the Cold War," in *Science and Technology in the Global Cold War*, edited by Oreskes and Krige, Cambridge, MA: MIT Press, 2014, 11–24.

10. For an overview: Ronald E. Doel, "Earth Sciences and Geophysics," in *Science in the 20th Century*, edited by John Krige and Dominique Pestre, Amsterdam: Harwood Academic, 2003, 391–417; John Cloud, ed., "Science in the Cold War," Special Issue, *Social Studies of Science* 33, no. 5 (2003); James Rodger Fleming, *Fixing the Sky: The Checkered History of Weather and Climate Control*, New York: Columbia University Press, 2010.

11. Simone Turchetti. "A Most Active Customer: How the U.S. Administration Helped the Italian Atomic Energy Project to 'De-Develop,'" *Historical Studies in the Natural Sciences* 44, no. 5 (2014): 470–502.

12. Interesting reflections on historiographical tensions in defense research on both sides of the Atlantic are in David Edgerton, *Warfare State: Britain, 1920–1970*, Cambridge, UK: Cambridge University Press, 2006, 324–26.

13. Key aspects of NATO's history are the focus of a series of lectures by historian and diplomat Jamie Shea (currently deputy assistant secretary general for NATO's Emerging Security Challenges division), arranged on the occasion of NATO's sixtieth anniversary (available at http://www.nato.int/60years

/video/videos-jamie.html). For a quick overview on NATO as an organization, see also Jennifer Medcalf, *NATO: A Beginner's Guide*, Oxford, UK: One World, 2005.

14. Lawrence Kaplan, *The United States and NATO: The Formative Years*, Lexington: University Press of Kentucky, 1984; Kaplan, *The Long Entanglement*, Westport, CT: Praeger, 1999. See also the essays in Gustav Schmidt, ed., *A History of NATO: The First Fifty Years*, Basingstoke, UK: Palgrave, 2001.

15. Ekavi Athanassopoulous, *Turkey–Anglo-American Security Interests, 1945–1952: The First Enlargement of NATO*, London: Frank Cass, 1999; Leopoldo Nuti, ed., *The Crisis of Détente in Europe: From Helsinki to Gorbachev, 1975–1985*, London: Routledge, 2008; Ronald E. Powaski, *The Entangling Alliance: The United States and European Security, 1950–1993*, Westport, CT: Greenwood, 1994; Linda Russo, *Propaganda and Intelligence in the Cold War: The NATO Information Service*, London: Routledge, 2014; David N. Schwartz, *NATO's Nuclear Dilemmas*, Washington, DC: Brookings Institution Press, 1983; Kristan Stoddart, *The Sword and the Shield: Britain, America, NATO, and Nuclear Weapons, 1970–1976*, New York: Palgrave, 2014.

16. Pledges for a wider use of science diplomacy in international relations are in Carlos Moedas (EU commissioner for research, science, and innovation), "Science Diplomacy in the European Union," *Science and Diplomacy*, 5, no. 1 (2016); and Vaughan C. Turekian and Norman P. Neureiter, "Science and Diplomacy: The Past as Prologue," *Science and Diplomacy*, 1, no. 1 (2012), http://www.sciencediplomacy.org/editorial/2012/science-and-diplomacy. See also Kristin M. Lord and Vaughan C. Turekian, "Time for a New Era of Science Diplomacy," *Science* 315 (2007): 769–70. On soft power, see Joseph Nye, *Soft Power: The Means to Success in World Politics*, New York: Public Affairs, 2004.

17. AAAS, "Science and Diplomacy: A Conceptual Framework, 2009," https://www.aaas.org/sites/default/files/scidip_framework_aaas_2009.pdf; AAAS and Royal Society, *New Frontiers in Science Diplomacy: Navigating the Changing Balance of Power*, London: Science Policy Centre, 2010, https://royalsociety.org/~/media/Royal_Society_Content/policy/publications/2010/4294969468.pdf. See also Frank L. Smith III, "Advancing Science Diplomacy: Indonesia and the US Naval Medical Research Unit," *Social Studies of Science* 44, no. 6 (2014): 825–47; Gabriella Paár-Jákli, *Networked Governance and Transatlantic Relations: Building Bridges through Science Diplomacy*, New York: Routledge, 2014.

18. AAAS and Royal Society, *New Frontiers*, 2 and 29. Jim McQuaid (NATO Science for Peace and Security) was a member of the advisory group designing the Royal Society's *New Frontiers in Science Diplomacy*. On the history of science diplomacy, see also Elisabeth Crawford, Terry Shinn, and Sverker Sörlin, *Denationalizing Science: The Context of International Scientific Practice*, Dordrecht, Neth.: Kluwer, 1993; John Krige and Kai-Henrik Barth, eds., *Global Power Knowledge: Science and Technology in International Affairs*, Chicago: University of Chicago Press, 2006; Sverker Sörlin, "Ice Diplomacy and Climate Change:

Hans Ahlmann and the Quest for a Nordic Region beyond Borders," in S.Sörlin, ed., *Science, Geopolitics and Culture in the Polar Region*, Farnham, UK: Ashgate, 2013, 23–55.

19. Marc Trachtenberg, *A Constructed Peace: The Making of the European Settlement, 1945–1963*, Princeton, NJ: Princeton University Press, 1999; Ralph Dietl, "In Defence of the West: General Lauris Norstad, NATO Nuclear Forces and Transatlantic Relations, 1956–1963," *Diplomacy and Statecraft* 17 (2006): 347–92.

20. In a way similar to what has happened in the recent history of Antarctica (also bringing together the promotion of science and transnational governance). See Simone Turchetti, Simon Naylor, Katrina Dean, and Martin Siegert, "On Thick Ice: Scientific Internationalism and Antarctic Affairs, 1957–1980," *History and Technology* 24, no. 4 (2008): 351–76.

21. On parallel diplomacy, see Louise Diamond and John W. McDonald, *Multi-Track Diplomacy: A Systems Guide and Analysis*, West Hartford, CT: Kumarian, 1996.

22. Trachtenberg, *Constructed Peace*, chapter 3.

23. I. I. Rabi to John McLucas, December 5, 1967, in Isidor Isaac Rabi Papers, US Library of Congress, Washington, DC [RABI from here on], box 39, folder 2.

24. These issues are admirably discussed in Michael D. Gordin, *Scientific Babel: How Science Was Done before and after Global English*, Chicago: University of Chicago Press, 2015.

25. Doing what Thomas Gieryn defines as "boundary work" at the edge of science- and policy-making. See Thomas F. Gieryn, *Cultural Boundaries of Science: Credibility on the Line*, Chicago: Chicago University Press, 1999. See also Gieryn, "Boundary Work and the Demarcation of Science from Non-science," *American Sociological Review* 48 (1983), 781–95.

26. Krige, *American Hegemony*, 7–8.

27. See, for instance, Lawrence S. Wittner, *Toward Nuclear Abolition*, Stanford, CA: Stanford University Press, 130–54; Rex Weyler, *Greenpeace: How a Group of Journalists, Ecologists, and Visionaries Changed the World*, Emmaus, PA: Rodale, 2004.

28. J. Brooks Flippen, "Richard Nixon, Russell Train, and the Birth of Modern American Environmental Diplomacy," *Diplomatic History* 32, no. 4 (2008): 613–38; Stephen J. Macekura, *Of Limits and Growth: The Rise of Global Sustainable Development in the Twentieth Century*, Cambridge, UK: Cambridge University Press, 2015, 17.

29. See, for instance, Kurkpatrick Dorsey, *The Dawn of Conservation Diplomacy: U.S.–Canadian Wildlife Protection Treaties in the Progressive Era*, Seattle: University of Washington Press, 1998; Toshihiro Higuchi, "Tipping the Scale of Justice: The Fallout Suit of 1958 and the Environmental Legal Dimension of Nuclear Pacifism," *Peace and Change* 38, no. 1 (2013): 33–55.

30. On environmental groups in the United States, see Christopher Bosso, *Environment, Inc.: From Grassroots to Beltway*, Lawrence: University Press of Kan-

sas, 2005. On those in Scandinavia, see Andrew Jamison, *The Making of Green Knowledge: Environmental Politics and Cultural Transformation*, Cambridge, UK: Cambridge University Press, 2001. On scientists as environmental campaigners, see David Zierler, *The Invention of Ecocide: Agent Orange, Vietnam, and the Scientists Who Changed the Way We Think about the Environment*, Athens: University of Georgia Press, 2011. See also Robert Gottlieb, *Environmentalism Unbound: Exploring New Pathways for Change*, Cambridge, MA: MIT Press, 2001.

31. Macekura has recently shown how environmental diplomacy helped the US administration to stave off investments in international cooperation in the Third World. Macekura, *Of Limits and Growth*, 5. See also on this Audra J. Wolfe, *Competing with the Soviets: Science, Technology, and the State in Cold War America*, Baltimore, MD: Johns Hopkins University Press, 2015, 60–73; Amanda Kay McVety, *Enlightened Aid: U.S. Development as Foreign Policy in Ethiopia*, Oxford: Oxford University Press, 2012.

32. Thomas Robertson refers to the instrumental uses of environmental protection as the "dark underbelly" of environmentalism. Robertson, *Malthusian Moment*, xiv.

33. "The participants labeled this dichotomy [scientific rationalism vs. countercultural beliefs and values] as 'mechanics versus mystics,' and it would remain a fundamental cleavage within the organization throughout the 1970s," Frank Zelko, *Make It a Green Peace! The Rise of Countercultural Environmentalism*, New York: Oxford University Press, 2013, 7.

34. See on this Stephen Bocking, *Nature's Experts: Science, Politics, and the Environment*, New Brunswick, NJ: Rutgers University Press, 2006, 38. See also Jamison, *Making of Green Knowledge*, 12.

35. Andrew J. Hoffman, *How Culture Shapes the Climate Change Debate*, Stanford, CA: Stanford University Press, 2015, 30; Mike Hulme, *Why We Disagree about Climate Change*, Cambridge, UK: Cambridge University Press, 2009, 230–31.

36. Naomi Oreskes, "Changing the Mission: From Cold War to Climate Change," in *Science and Technology in the Global Cold War*, edited by Oreskes and Krige, 141–88. See also T. C. Vance and R. E. Doel, "Graphical Methods and Cold War Scientific Practice: The Stommel Diagram's Intriguing Journey from the Physical to the Biological Environmental Sciences," *Historical Studies in the Natural Sciences* 40 (2010): 1–47; Sebastian Grevsmühl, "Serendipitous Outcomes in Space History: From Space Photography to Environmental Surveillance," in *Surveillance Imperative*, edited by Turchetti and Roberts, 171–94.

37. Some ongoing efforts are described on the NATO website: www.nato.int/cps /en/natohq/news_116161.htm.

38. I used a similar method in completing my previous monograph *The Pontecorvo Affair: A Cold War Defection and Nuclear Physics*, Chicago: University of Chicago Press, 2012.

39. One could, for instance, liken NATO to many other Cold War organizations defined by distinctive power and gender relations. See on this Carol Cohn

in "Sex and Death in the Rational World of Defense Intellectuals," reprinted in Mary Wyer, Mary Barbercheck, Donna Cookmeyer, Hatice Ozturk, and Marta Wayne, eds., *Women, Science and Technology*, London: Routledge, 2001, 99–116.

40. On transnational history, see Pierre Y. Saunier and Akira Iriye, eds., *The Palgrave Dictionary of Transnational History*, New York: Palgrave, 2009. See also Ian Tyrrell, "What Is Transnational History?" Excerpt of a paper given at the *École des hautes études en sciences sociale* (Paris) in January 2007, http://iantyrrell.wordpress.com/what-is-transnational-history. See also Simone Turchetti, Néstor Herran, and Soraya Boudia, "Introduction: Have We Ever Been 'Transnational'? Towards a History of Science across and beyond Borders," *British Journal for the History of Science* 45 (2012): 319–36. John Krige describes these encounters between national groups as defining processes of "mongrelization"—the creation of hybrid forms of transnational knowledge. John Krige, "Hybrid Knowledge: The Trans-National Co-production of the Gas Centrifuge for Uranium Enrichment in the 1960s," *British Journal for the History of Science* 45 (2012): 337–57.

CHAPTER 1

1. "House of Commons," *Times* (London), June 14, 1945.
2. On war (including World War II) and the environment, see John R. McNeill, *Something New under the Sun: An Environmental History of the Twentieth-Century World*, London: Penguin, 2000, 341–48.
3. Such as the International Union for Conservation of Nature (IUCN). Max Nicholson, *The Environmental Revolution: A Guide for the New Masters of the World*, Harmondsworth, UK: Penguin, 1970, 31. See also Macekura, *Limits and Growth*, 39; Bosso, *Environment, Inc.*, 33.
4. Raf de Bont and Geert Vanpaemel, "The Scientist as Activist: Biology and the Nature Protection Movement, 1900–1950," *Environment and History* 18 (2012), 203–8.
5. Nicholson wrote in 1970, "until quite recently its common use was largely topographical; it signified the surroundings, loosely indicated, of some point or area as a city," *Environmental Revolution*, 278.
6. On this, see James R. Fleming, "Sverre Pettersen, the Bergen School and the Forecasts for D-Day," *Proceedings of the International Commission on History of Meteorology* 1, no. 1 (2004): 75–83.
7. NATO Information Service, *Facts about the North Atlantic Treaty Organization*, Utrecht: NATO/Bosch, 1962, 15–18.
8. Kaplan, *Long Entanglement*, 2–3.
9. Prominent examples are the 1778 Franco-American Treaty and World War II's United Nations. Notably, US participation in World War I was framed in terms of contribution by an associated power rather than an ally. See Jamie Shea, "NATO's Anxious Birth," https://www.nato.int/cps/en/natohq

/opinions_139301.htm. See also Powaski, *Entangling Alliance*, xvii; Trachtenberg, *Constructed Peace*, 119.

10. Trachtenberg, *Constructed Peace*, 109–12.

11. Through the 1951 Mutual Security Act. See on this McVety, *Enlightened Aid*, 88 and 108–12. On the reorganization of defense in the United States, see Kevles, "Cold War, Hot Physics," 239–64.

12. Cited in Sean M. Maloney, *Securing Command of the Sea: NATO Naval Planning, 1948–1954*, Annapolis, MD: Naval Institute Press, 1995, 135. See also Martin A. Smith, "Britain Nuclear Weapons and NATO in the Cold War and Beyond," *International Affairs* 87, no. 6 (2011): 1385–99.

13. Athanassopoulous, *Turkey–Anglo-American Security Interests*, 203–16. British views on defense planning differed significantly, as the Mediterranean was considered to be a separate defense entity from Western Europe. See Effie G. H. Pedaliu, *Britain, Italy and the Origins of the Cold War*, New York: Palgrave, 2003, 120–27.

14. Trachtenberg, *Constructed Peace*, 158–59. See also 172–73 on the actual interpretation of this doctrine. See further Michael O. Wheeler, "NATO Nuclear Strategy, 1949–1990," in *History of NATO*, edited by Schmidt, 3: 121–40.

15. Cristopher J. Bright, *Continental Defense in the Eisenhower Era: Nuclear Antiaircraft Arms and the Cold War*, New York: Palgrave, 2010, 1.

16. Kaplan, *Long Entanglement*, 62.

17. Trachtenberg, *Constructed Peace*, 122–23.

18. For the Italian case, see Leopoldo Nuti, *La Sfida Nucleare: La Politica Estera Italiana e le Armi Atomiche, 1945–1991*, Bologna, Italy: Mulino, 2007.

19. Rolf Tamnes, "The Strategic Importance of the High North during the Cold War," in *History of NATO*, edited by Schmidt, 3: 257–74; and Nikolaj Petersen, "Denmark's Fifty Years with NATO," 275–94, also in *History of NATO*, edited by Schmidt, 3: 275–94.

20. Lino Camprubí and Sam Robinson, "A Gateway to Oceanic Circulation: Submarine Surveillance and the Contested Sovereignty of Gibraltar," *Historical Studies in the Natural Sciences*, 46, no. 4 (2016): 429–59, on 442.

21. Mikael af Malmborg, "Sweden—NATO's Neutral 'Ally'? A Post-revisionist Account," in *History of NATO*, edited by Schmidt, 3: 295–316. See also Hans Weinberger, "The Neutrality Flagpole: Swedish Neutrality Policy and Technological Alliances, 1945–1970," in *Technologies of Power: Essays in Honor of Thomas Parker Hughes*, edited by Michael Thad Allen and Gabrielle Hecht, Cambridge, MA: MIT Press, 2001, 295–332.

22. Mikael Nilsson, "Amber Nine: NATO's Secret Use of a Flight Path over Sweden and the Incorporation of Sweden in NATO's Infrastructure," *Journal of Contemporary History* 44, no. 2 (2009): 287–307.

23. Matthias Heymann, Henry Nielsen, Kristian Hvidtfelt Nielsen, and Henrik Knudsen, "Small State versus Superpower: Science and Geopolitics in Greenland in the Early Cold War," in *Cold War Science and the Transnational Circulation of Knowledge*, edited by van Dongen, Hoeneveld, and Streefland, 243–71,

on 269; Janet Martin-Nielsen, *Eismitte in the Scientific Imagination: Knowledge and Politics at the Center of Greenland*, New York: Palgrave Macmillan, 2013, 81. See also Ronald E. Doel, Kristine C. Harper, and Matthias Heymann, eds., *Exploring Greenland: Cold War Science and Technology on Ice*, New York: Palgrave, 2016.

24. John Krige, "Atoms for Peace, Scientific Internationalism, and Scientific Intelligence," in *Global Power Knowledge*, edited by Krige and Barth, 161–81, on 163.

25. Within the framework of the Western European Union. See Dietl, "In Defence of the West," 352–53.

26. Turchetti, *Pontecorvo Affair*, 157–58.

27. AGARD Secretariat, NATO-AGARD, 1952, 1–6; copy in Theodore von Kármán papers, California Institute of Technology Archive, Pasadena, CA [TVK from here on], box 34, folder 5.

28. Van der Bliek, ed., *AGARD*, 3–1.

29. Ibid., 2–1.

30. See Michael H. Gorn, *The Universal Man: Theodore von Kármán's Life in Aeronautics*, Washington, DC: Smithsonian Institution Press, 1992.

31. T. von Kármán to R. Lovett, December 19, 1950; copy in RABI, box 39, folder 6.

32. McVety, *Enlightened Aid*, 112.

33. DOD Research and Engineering, "Guide to MWDP," September 23, 1959, 1; copy in Howard Percy Robertson Papers, California Institute of Technology, Pasadena, California [ROB from here on], box 15, folder 9. On the Mutual Defense Assistance Programs, see Lorenza Sebesta, "American Military Aid and European Rearmament," in *NATO: The Founding of the Atlantic Alliance and the Integration of Europe*, edited by Francis H. Heller and John R. Gillingham, Basingstoke, UK: Macmillan, 1992, 283–310; Jacqueline McGlade, "NATO Procurement and the Revival of European Defense," in *History of NATO*, edited by Schmidt, 3: 13–28.

34. Defence Research Directors, Summary Record of Meeting, January 21, 1959, Secret, NATO Archives, NATO Headquarters, Brussels, Belgium [NATO from here on], AC/137(DR)R/1, 41. See also C. Güner Omay, *Prof. Dr. Egbert Adriaan Kreiken: Founder of the Ankara University Observatory and a Volunteer of Education*, Birinci Basım, Turkey: Kasım, 2011, 12–13.

35. Defence Research Directors, Summary Record of Meeting, October 8, 1959, Secret, NATO, AC/137(DR)R/2, 26–27.

36. Wolfe, *Competing with the Soviets*, 32–33. See also Jeroen van Dongen and Friso Hoeneveld, "*Quid pro Quo*: Dutch Defense Research during the Early Cold War," in *Cold War Science and the Transnational Circulation of Knowledge*, edited by van Dongen, Hoeneveld and Streefland, 101–21, on 111–12.

37. Jesse L. Greenstein, "Howard Percy Robertson, January 27, 1903–August 26, 1961," *NAS Biographical Memoirs* 51 (1980): 343–426, on 346. These activities include his work as main referee for the journal *Physical Review*. On this, see

Roberto Lalli, "'Dirty Work,' but Someone Has to Do It: Howard P. Robertson and the Refereeing Practices of *Physical Review* in the 1930s," *Notes and Records of the Royal Society of London* 70, no. 2 (2016): 151–74.

38. Caltech News Bureau press release, May 8, 1954, in TVK, box 25, folder 4; H. P. Robertson, "The Impact of Science on Military Thought," unpublished manuscript, ROB, box 18, folder 12. On the origins of operations research: Krige, *American Hegemony*, chapter 8; Erik P. Rau, "Technological Systems, Expertise, and Policy Making: The British Origins of Operational Research," in *Technologies of Power: Essays in Honor of Thomas Parker Hughes*, edited by Michael Thad Allen and Gabrielle Hecht, Cambridge, MA: MIT Press, 2001, 215–52. Erickson et al., *How Reason Almost Lost Its Mind*, 51–80. On Robertson's activities, see also William Thomas, *Rational Action: The Sciences of Policy in Britain and America, 1940–1960*, Cambridge, MA: MIT Press, 2015, 106 and 112.

39. Kevles, "Cold War, Hot Physics," 252.

40. H. P. Robertson to Joyce Hannaum, ROB, box 11, folder 4.

41. H. P. Robertson, "The Impact of Science on Military Thought," ROB, box 18, folder 12.

42. Ibid., 19. On scientific intelligence, see Ronald E. Doel and Alan A. Needell, "Science, Scientists, and the CIA: Balancing International Ideals, National Needs, and Professional Opportunities," *Intelligence and National Security* 12, no. 1 (1997): 59–81; Reginald Victor Jones, in *Most Secret War* (London: Penguin, 2009), discusses British contributions.

43. Raymond H. Dawson and George E. Nicholson, "NATO and the SHAPE Technical Center," *International Organization*, 21, no. 3 (1967): 565–91, on 568.

44. Ibid.

45. "By the early Cold War, military patrons declared that military applications existed for each component branch of the earth sciences." Doel, "Earth Sciences and Geophysics," 402.

46. On this, see Kevles, "Cold War, Hot Physics," 245. See also Jacob Darwin Hamblin, *Oceanographers and the Cold War: Disciples of Marine Science*, Seattle: University of Washington Press, 2005; Eric L. Mills, *The Fluid Envelope of Our Planet: How the Study of Ocean Currents Became a Science*, Toronto: University of Toronto Press, 2011.

47. On the origins of the IGY, see Fae L. Korsmo, "The Birth of the International Geophysical Year," *Leading Edge* 10 (2007):1312–16; Walter Sullivan, *Assault on the Unknown: The International Geophysical Year*, New York: McGraw-Hill, 1961. On the politics of IGY data, see Elena Aronova, Karen Baker, and Naomi Oreskes. "Big Science and Big Data in Biology: From the International Geophysical Year through the International Biological Program to the Long-Term Ecological Research Program, 1957–Present," *Historical Studies in the Natural Sciences* 40, no. 2 (2010): 183–224.

48. Alan Needell, *Science, Cold War and the American State: Lloyd V. Berkner and the Balance of Professional Ideals*, Amsterdam: Harwood Academic, 2000, 313–14.

49. AGARD Secretariat, "NATO-AGARD," 1952, 1–6; copy in TVK, box 34, folder 5.

50. Harry Wexler, US Weather Bureau, to T. von Kármán, December 21, 1954, in TVK, box 34. The proceedings were published as Reginald C. Sutcliffe, ed., *Polar Atmosphere Symposium*, part I, *Meteorology Section*, New York: Pergamon, 1958. On Wexler, see also James Rodger Fleming, *Inventing the Atmospheric Science: Bjernknes, Rossby, Wexler, and the Foundations of Modern Meteorology*, Cambridge, MA: MIT Press, 2016, chapter 4.

51. On US research efforts predating the mission, see Ronald E. Doel, "Defending the North American Continent: Why the Physical Environmental Sciences Mattered in Cold War Greenland," in *Exploring Greenland*, edited by Doel, Harper, and Heymann, 25–46. See also Martin-Nielsen, *Eismitte in the Scientific Imagination*, 66.

52. "Defensively the Arctic has assumed major importance with respect to the Distant Early Warning Line and our need for extending our early warning and detection system as far toward the USSR as operationally possible." US Joint Chiefs of Staff document cited in Tamnes, "Strategic Importance," 257–74, on 259–60.

53. F. L. Wattendorf, "Opening Address," in *Polar Atmosphere Symposium*, edited by Sutcliffe, viii.

54. Charles C. Bates, "Current Status of Sea Ice Reconnaissance and Forecasting for the American Arctic," in *Polar Atmosphere Symposium*, edited by Sutcliffe, 285–322. Ice forecasting united "the drive to know the earth through observation and calculation and to know the enemy and its capabilities through contact with its scientists," wrote Peder Roberts in "Scientists and Sea Ice Surveillance in the Early Cold War," in *Surveillance Imperative*, edited by Turchetti and Roberts, 125.

55. K. Weekes, ed., *Polar Atmosphere Symposium*, part II, *Ionospheric Section*, New York: Pergamon, 1957.

56. USAF Lt. Col. Orlo F. Duker (NATO Standing Group Deputy Secretary), "Numerical Weather Prediction in NATO," December 29, 1959, SGM-79–57, NATO.

57. Avionics Panel newsletter 1, February 14, 1958; copy in TVK, box 35, folder 1.

58. As documented in Fleming, *Inventing Atmospheric Science*, especially 13–76.

59. Interview by the author with Nils Holme and Olav Blichner, Oslo, June 3, 2014.

60. Rankin, *After the Map*, 243.

61. NATO Information Service, *Facts about the North Atlantic Treaty Organization*, 131–36. Technical details about the system are discussed in Joint Technical Advisory Committee (JTAC), "Radio Transmission by Ionospheric and Tropospheric Scatter," *IRE Proceedings*, January 1960, 4–44. On the telephone network, see Sanne Aagard Jensen, "Connecting the Alliance: Communications Infrastructure on the NATO Agenda in the 1950s," in *Historicizing Infrastructure*, edited by Andreas Marklund and Mogens Rüdiger, Copenhagen: Aalborg University Press, 2017, 183–211. The geographical layout of the ACE

High network can be seen at this webpage: http://www.ace-high-journal.eu
/technik--handbuch-vi-.html.

62. On the DEW line, see Paul Edwards, *The Closed World: Computers and the Politics of Discourse in Cold War America*, Cambridge, MA: MIT Press, 1996, chapter 3.

63. AGARD publications listed in R. J. Lees, "Avionics Panel," 1959; copy in TVK, box 35, folder 1.

64. DOD Research and Engineering, "Guide to MWDP," September 23, 1959, 1, ROB, box 15, folder 9. The E layer is 90 to 150 km above the earth surface, the F layer 150 to 500 km.

65. MWDP Project Proposal: Optical-Radio Observatory (Turkey), October 3, 1958, in ROB, box 15, folder 7. On February 18, 1959, General Larkin attended its inauguration ceremony. Fuat Uluğ, presentation at the 1959 AGARD meeting, in TVK, box 34, folder 11.

66. Defence Research Directors Meeting, October 8, 1959, Secret, NATO, AC/137-(DR)R/2, 38.

67. I. Ranzi and Pietro Dominici, "Backscatter Sounding during Ionospheric Storms," *Scientific Note* 5, May 15, 1961, EOARDC No. AF(61(052)—139, www
.dtic.mil/dtic/tr/fulltext/u2/268084.pdf.

68. Van der Bliek, ed., *AGARD*, 2–3.

69. Russo, *Propaganda and Intelligence*, 62–66. On the CIA's patronage, see Frances Stonor Sounders, *The Cultural Cold War*, New York: New Press, 1999.

70. H. P. Robertson to J. Wheeler, July 22, 1957, Koepfli Papers, California Institute of Technology Archive, Pasadena, California [KOE from here on], box 1, folder 5.

71. Turchetti, "Most Active Customer," 470–502.

72. On its history, see Armin Hermann, John Krige, Ulrike Mersits, Dominique Pestre, Laura Weiss, and Lanfranco Belloni, *History of CERN*, 3 vols., Amsterdam: North-Holland, 2000.

73. Krige, *American Hegemony*, 257.

74. "Science and Foreign Relations: Berkner Report to the U.S. Department of State," *Bulletin of the Atomic Scientists* 6, no. 10 (1950), 293–98. Strong was one of Robertson's CIA associates and the chemist Koepfli was one of his colleagues at Caltech (and a close friend too). H. Rudolph to J. Koepfli, September 9, 1949, in KOE, box 1, folder 2. On Berkner, see also Needell, *Science, Cold War and the American State*.

75. Peter Galison and Barton Bernstein, "In Any Light: Scientists and the Decision to Build the Superbomb," *Historical Studies in the Physical and Biological Sciences* 19 (1989): 267–347. See also Paul Forman, "Inventing the Maser in Postwar America," *Osiris* 7 (1992): 105–34; John S. Rigden, *Rabi: Scientist and Citizen*, Cambridge, MA: Harvard University Press, 1987.

76. John Krige, "I. I. Rabi and the Birth of CERN," *Physics in Perspective* 7 (2005): 150–64, on 162.

77. Michael A. Day, "Oppenheimer and Rabi: American Cold War Physicists as

Public Intellectuals," in *The Atomic Bomb and American Society: New Perspectives*, edited by Rosemary B. Mariner and G. Kurt Piehler, Knoxville: University of Tennessee Press, 2009, 307–28.

78. Rigden, *Rabi*, 56–58, and 266–67.

79. Nicholas DeWitt, *Soviet Professional Manpower: Training and Supply*, Washington, DC: National Science Foundation, 1955. See also David Kaiser, "The Physics of Spin: Sputnik Politics and American Physicists in the 1950s," *Social Research* 73 (2006): 1225–52.

80. See on this Mary Jo Nye, *Michael Polanyi and His Generation: Origins of the Social Construction of Science*, Chicago: University of Chicago Press, 2011, 211.

81. Russo, *Propaganda and Intelligence*, 227.

82. On NATO's recent commemoration of the report, see http://www.nato.int /cps/en/natohq/news_139354.htm. Not to be confused with another Three Wise Men (Louis Armand, Franz Etzel, and Francesco Giordani) who issued the report "A Target for EURATOM" the following year.

83. Terms of Reference for the Three Wise Men were set in the NAC meeting of May 5, 1956 (NAC Summary Record of Meeting, May 5, 1956, NATO, C-R(56)23, Secret, 6. "Report of the Committee of Three on Non-military Co-operation in NATO," December 13, 1956, Item IV, NATO, C-M(56)127(Revised), Unclassified, 19. Available at www.nato.int/cps/en/natohq/official_texts _17481.htm. See also Russo, *Propaganda and Intelligence*, 77–78.

84. "Report of the Committee of Three," Item IV, Point 69.

85. "Further Action by NATO in the Field of Scientific and Technical Co-operation," April 1, 1957, NATO, C-M(57)50, 1.

86. Second note by the US representative (at the Working Group to Consider Further Action by NATO in the Field of Scientific and Technical Co-operation), March 13, 1957, Confidential, NATO, AC/123-WP/3.

87. Robertson to Wheeler, July 22, 1957, in KOE, box 1, folder 5. See also Krige, *American Hegemony*, 201–4. Wheeler's committee included prominent US science administrators such as National Academy of Sciences President Detlev Bronk and MIT President James Killian.

88. Summary Record of NAC Meeting, April 29, 1957, C-R(57)25, 8.

89. Koepfli argued that he and Robertson were "very, very, very dear friends" in his interview with Ron Doel, August 3, 1995, https://www.aip.org/history -programs/niels-bohr-library/oral-histories/31375. By then, Koepfli also featured in a Science and Freedom meeting organized in Hamburg by the Congress for Cultural Freedom. Nye, *Michael Polanyi and His Generation*, 211.

90. J. G. P. Spicer, "Scientific Co-operation in NATO: General Approach," June 20, 1957; copy in Solly Zuckerman Archive, University of East Anglia, Norwich, UK [ZUC from here on], SZ/NATO/1.

91. Powaski, *Entangling Alliance*, 45.

92. See on this Turchetti, "'In God We Trust, All Others We Monitor': Seismol-

ogy, Surveillance, and the Test Ban Negotiations," in *Surveillance Imperative*, edited by Turchetti and Roberts, 85–102.

93. "Further Action by NATO in the Field of Scientific and Technical Co-operation: Report to Council by the Task Force," November 4, 1957, NATO, C-M(57)130.
94. Ibid.
95. NATIS, *Facts about the North Atlantic Treaty Organization*, 282.
96. Ibid., 6.
97. On Zuckerman, see John Peyton, *Solly Zuckerman: A Scientist Out of the Ordinary*, London: John Murray, 2001; Peter Leslie Krohn, "Solly Zuckerman, 30 May 1904–1 April 1993," *Biographical Memoirs of Fellows of the Royal Society* 41 (1995): 575–98. Zuckerman discussed his life in Zuckerman, *Monkeys, Men, and Missiles: An Autobiography, 1946–88*, New York: Norton, 1989. On his war work, see also Thomas, *Rational Action*, 83–84.
98. Krohn, "Solly Zuckerman," 588.
99. H. Robertson to S. Zuckerman, April 21, 1960, ROB, box 6, folder 27.
100. BJSM, Washington to MoD, London, Restricted, June 24, 1957, ZUC, SZ/CSA/126.
101. H. Robertson to S. Zuckerman, January 7, 1960, ZUC, SZ/CSA/126.
102. David Wineland, "Norman Ramsey (1915–2011)," *Nature* 480 (2011), 182.
103. A. B. Clark, NSA, to H. Robertson, February 11, 1954, in ROB, box 18, folder 13.
104. Ramsey to Zuckerman, March 1, 1958, in ZUC, SZ/NATO/3.
105. Note by the US Delegation, August 21, 1958, in RABI, box 38, folder 10. Krige, *American Hegemony*, 244–45. Another ad hoc group of specialists was set up to organize initiatives on human factors (or defense psychology).
106. Science Committee, Summary Record of Meeting, April 25, 1958, NATO, AC/137-R/1, 5.
107. Major General Theodore W. Parker, Invited Statement on Matters of Defence Science, May 6, 1958, Secret, NATO, AC/137-D/10.
108. Ibid., 30.
109. Science Committee, Summary Record of Meeting, April 25, 1958, NATO, AC/137-R/1, 31.
110. See on this Turchetti, "Most Active Customer," 470–502, on 490–93; Krige, *Sharing Knowledge, Shaping Europe*, chapter 2 (especially 67–74); Grégoire Mallard, "L'Europe Puissance Nucléaire, Cet Obscur Objet du Désir," *Critique Internationale* 42 (2009): 141–63.
111. Science Committee, Summary Record of Meeting, January 19, 1959, Secret, NATO, AC/137-R/3, 25–26; Science Committee, Report by the Working Party on Declassification of Scientific Information, July 4, 1958, Restricted, NATO, AC/137-D/17. An account of the declassification strategies adopted within the committee is in Simone Turchetti, "A Need-to-Know-More Criterion? Science and Information Security at NATO during the Cold War," in *Cold War Science*

and the Transatlantic Circulation of Knowledge, edited by Jeroen Van Dongen, Friso Hoeneveld, and Abel Streefland, 36–58.

112. William Nierenberg, NATO Science Activities, March 4, 1964, in RABI, box 38, folder 10, 6.

CHAPTER 2

1. Manlio Brosio, NATO Secretary General, in speeches commemorating the tenth anniversary of the foundation of the NATO Science Committee, March 21, 1968, NATO, AC/137-D/330.
2. DRD, Subjects of Military Importance Which Have a Large Element of Scientific Research, November 28, 1958, Restricted, NATO, AC/137(DR)WP/1.
3. Ibid., 14.
4. The budget available for the Advanced Study Institute also grew but continued to be lower. Budget data (in US dollars) are from "Annual Expenditures on Scientific Programmes" in Scientific Affairs Division, *NATO and Science,* 18.
5. The program drew on a preexisting program sponsored by the US administration (the Foreign Leader Program) targeting humanists. Russo, *Propaganda and Intelligence,* 201–2.
6. Science Committee, Summary Record of Meeting, April 25, 1958, NATO, AC/137-R/1.
7. Science Committee, Summary Record of Meeting, August 6, 1964, NATO, AC/137-R/19, 4–8. Data (in number of NATO fellowships per discipline) are from "Subjects Studied by Fellows during the Years 1959–1963" in NATO Scientific Affairs Division, *NATO and Science,* 30.
8. Science Committee, Summary Record of Meeting, August 6, 1964, NATO, AC/137-R/19, 8–10.
9. CIA, "Development of Nuclear Capabilities by Four Countries: Likelihood and Consequences," NIE 100-2-58, July 1, 1958, 12, https://www.cia.gov /library/readingroom/docs/DOC_0001108555.pdf. For an overview on US surveillance activities on the French nuclear program, see Jeffrey Richelson, *Spying on the Bomb: American Nuclear Intelligence from Nazi Germany to Iran and North Korea,* New York: Norton, 2006.
10. Proposal by the French government for a Western foundation for scientific research, Secret, NATO, AC/137-D/2. On Danjon, see also Elisabeth Crawford and Josiane Olff-Nathan, eds., *La Science sous Influence: L'Université de Strasbourg Enjeu des Conflits Franco-Allemands, 1972–1945,* Strasbourg: Nuée Bleue, 2005.
11. Dominique Pestre, "Louis Néel et le Magnétisme à Grenoble," *Cahiers pour l'Histoire du CNRS* 8 (1990), 101, http://www.histcnrs.fr/cahiers-cnrs/pestre .pdf.
12. Ad Hoc Working Party on Western Science Foundation, April 15, 1958, NATO Restricted, AC/137(WF)-N/1 (copy in ZUC, box SZ/NATO/1).
13. Note by the UK representative, February 17, 1959, in ZUC, SZ/NATO/1.

14. Science Committee, Summary Record of Meeting, January 19, 1959, Secret, NATO, AC/137-R/2.

15. Dominique Pestre, "Louis Néel," 52–54. See also Guthleben, *Histoire du CNRS de 1939 à Nos Jours*, Paris: Armand Colin, 2009, 126–27.

16. T. Sargeaunt, Defence Science, April 22, 1959, NATO, AC/137-WP/2 (copy in ZUC, SZ/NATO/2).

17. DRD, Summary Record of Meeting, January 21, 1959, Confidential, NATO, AC/137(DR)R/1.

18. Ibid., 14 and 23.

19. Ibid., 38.

20. Ibid., 44. On Brundrett, see Jon Agar and Brian Balmer, "British Scientists and the Cold War: The Defence Research Policy Committee and Information Networks, 1947–1963," *Historical Studies in the Physical and Biological Sciences* 28, no. 2 (1998): 209–52.

21. DRD, Summary Record of Meeting, January 21, 1959, Confidential, NATO, AC/137(DR)R/1, 50–54.

22. Down in the list were also communication for the services, chemistry of solid propellants, tactical mobility for the army, solid-state physics, anti-ballistic missile defense, service uses for electronic computers, materials research and testing, defense against low-level air attack. See "Defence Science," note by the deputy science adviser, April 22, 1959, Confidential, NATO, AC/137-WP/2.

23. Annex A to Science Committee, Summary Record of Meeting, January 19, 1959, Restricted, NATO, AC/137-R/3, 25–26.

24. Science Committee, Summary Record of Meeting, January 19, 1959, Restricted, NATO, AC/137-R/3, 42–43.

25. R. W. B. (Otto) Clarke, Treasury, to S. Zuckerman, January 2, 1958, ZUC, SZ/NATO/3. Otto was father to the Labour politician and former home secretary Charles Clarke (2004–2006 in Tony Blair's cabinet).

26. He wrote to Clarke, "I made the usual mumble about not being able to commit financially." S. Zuckerman to O. Clarke, May 1, 1959, in ZUC, SZ/NATO/3.

27. ACSP Overseas Scientific Relations Committee, NATO Research Grants Programme—note by the joint secretary, March 17, 1959; copy in ZUC, SZ/ACSP/11.

28. L. V. Berkner to J. Koepfli, August 10, 1959, and J. Koepfli to L. V. Berkner, August 24, 1959, in KOE, box 1, folder 6.

29. Von Kármán to Friedricks Kurt, New York University, November 12, 1957, in TVK, box 34, folder 20. The Italian physicist Edoardo Amaldi had also set forward a similar plan. See Lorenza Sebesta, "Italian Space Policy between Internal Renovation and External Challenges at the Turn of the Decade, 1957–1963," in *Italy in Space*, edited by de Maria and Orlando, 47–76, on 51. See also Lodovica Clavarino, *Scienza e Politica nell'Era Nucleare: La Scelta Pacifista di Edoardo Amaldi*, Rome: Carocci, 2014.

30. Needell, *Science, Cold War and the American State*, 325.

31. Science Committee, Summary Record of Meeting, January 19, 1959, NATO, AC/137-R/3, 20–21.

32. R. N. Quirk, "Outer Space Research," UK Cabinet memo, April 20, 1959; copy in ZUC, box SZ/NATO/2. On the Blue Streak cancelation, see Powaski, *Entangling Alliance*, 45.

33. Von Kármán to F. Wattendorf, July 22, 1959, in TVK, box 35, folder 4.

34. Advisory Group on Space Research, Summary Record of Meeting, March 22, 1960, AC/137(SR)R/1, 11–14.

35. See Michelangelo de Maria, "The San Marco Project (1961–1967)," in *Italy in Space*, edited by de Maria and Orlando, 77–106.

36. Sebesta, "Italian Space Policy," 51–52. Walter McDougall aptly notices that US-Europe collaborations in the manufacturing of space boosters was typified by "cooperation in science, aloofness in engineering." Walter A. McDougall, . . . *The Heavens and the Earth. A Political History of the Space Age*, New York: Basic Books, 1985, 352. See also John Krige, "Integration and the Non-Proliferation of Ballistic Missiles: The United States, the United Kingdom, and ELDO, 1966" in Krige, *Sharing Knowledge, Shaping Europe*, on 99–101.

37. Rankin, *After the Map*, 240–45. See also Andrew J. Butrica, ed., *Beyond the Ionosphere. Fifty Years of Satellite Communication*, Washington, DC: NASA, 1997.

38. On Vassy, see Dominique Pestre, "Studies of the Ionosphere and Forecasts for Radiocommunications: Physicists and Engineers, the Military and National Laboratories in France (and Germany) after 1945," *History and Technology* 13, no. 3 (1997): 183–205.

39. Communication by moon relay started at the US Naval Research Laboratory's Radar Division and expanded rapidly in the United States and Europe. David K. van Keuren, "Moon in Their Eyes: Moon Communication Relay at the Naval Research Laboratory, 1951–1962," in Butrica, ed., *Beyond the Ionosphere*, 9–18. See also Jon Agar, "Moon Relay Experiments at Jodrell Bank," in *Beyond the Ionosphere*, edited by Butrica, 19–30; Michael Mendillo, "Jules Aarons (1921–2008): Space Weather Pioneer," *Space Weather Quarterly*, 6, no. 2 (2009): 6–7; Sunanda Basu, "In Memoriam: Jules Aarons," *IEEE Antennas and Propagation Magazine* 51, no. 1 (2009), 137–38.

40. Jon Agar, *Science and Spectacle: The World of Jodrell Bank in Post-war British Culture* (Amsterdam: Harwood Academic, 1998); Bernard Lovell, *Out of the Zenith: Jodrell Bank, 1957–1970* (New York: Harper and Row, 1973).

41. See on this Aitor Anduaga, *Geophysics, Realism and Industry* (Oxford: Oxford University Press, 2016), 55–57.

42. Franco Samoggia, *Nello Carrara*, Siena, Italy: Carlo Cambi, 2006, 17–18 and 59–65.

43. Ruth P. Liebowitz, *Chronology: From the Cambridge Field Station to the Air Force Geophysics Laboratory, 1945–1985*, Hanscom, MA: Air Force Geophysics Laboratory, 1985, 18.

44. USAF, contract memorandum, August 6, 1958, Jodrell Bank Archive, Univer-

sity of Manchester Library [JBA from here on], folder "USAF Satellite & Moon Experiment," CS7/33/4.

45. Given the significance of this international collaborative exercise, I am reporting the author's list as it appears on the publication: J. Aarons and H. E. Whitney (Air Force Research Division, US), R. S. Roger and J. Thomson (Jodrell Bank, England), J. Bournazel and E. Vassy (Laboratory of Atmospheric Physics, France), H. A. Hess and K. Rawer (Ionospheric Institute of Bresiach, West Germany), N. Carrara and P. Checcacci (Centro Microonde, Italy), B. Landmark and J. Tröim (NDRE, Norway), B. Hultqvist, and L. Liszka (Kiruna Geophysical Observatory, Sweden), "Atmospheric Phenomena Noted in Simultaneous Observations of 1958 δ2 (Sputnik III)," *Planetary and Space Science* 5 (1961): 169–84.

46. The Véronique project was under the auspices of the French Service Prévision Ionosphérique de la Marine (SPIM). Pestre, "Studies of the Ionosphere," 187 and 198.

47. Nello Carrara, "Research Proposal on Propagation Characteristics with Satellites," September 1961, section 2.1, in NATO Surveillance Satellite, JBA, CS7/38/1.

48. Scientific Affairs Division, *NATO and Science*, 36.

49. Scientific Affairs Division, *NATO and Science*, appendix IV, Projects Involving International Co-operation.

50. Nello Carrara, Research Proposal on Propagation Characteristics with Satellites, September 1961, section 2.1, in NATO Surveillance Satellite, JBA, CS7/38/1.

51. Ibid.

52. Jules Aarons, John A. Klobuchar, Robin Stafford Allen, and Donald A. Guidice, "Second Symposium on Radio Astronomical and Satellite Studies of the Atmosphere," *IEEE Spectrum*, March 1966, 174–80.

53. Nello Carrara, Research Proposal for Second Year of Joint Satellite Propagation Studies, September 1962, in NATO Surveillance Satellite, JBA, CS7/38/1.

54. As he reported to von Kármán. L. Broglio, Scuola di Ingegneria Aeronautica, Rome, to T. von Kármán, AGARD, April 24, 1958, TVK, box 35, folder 1.

55. Enrico Medi to F. Seitz, April 1960, and E. Medi to P. Caldirola, July 21, 1960, in Istituto Nazionale di Geofisica Archive [ING from here on], Istituto Nazionale di Geofisica e Vulcanologia, Rome, box 41.

56. Scientific Affairs Division, *NATO and Science*, appendix IV, Projects Involving International Co-operation. See also van der Bliek, ed., *AGARD*, A5–7.

57. Scientific Affairs Division, *NATO and Science*, appendix IV, Projects Involving International Co-operation.

58. Scientific Affairs Division, *NATO and Science*, appendix IV, Table 2, Projects Not Necessarily Involving International Collaboration.

59. Michael Anastassiades, D. Ilias, G. Moraitis, and P. Giouleas, "Observations Made by the Ionospheric Institute of Athens during the Series of Nuclear

Weapon Tests at Novaja Zemlya between 10 September and 4 November 1961," in *Arctic Communications. Proceedings of the Eight Meeting of the AGARD Ionospheric Research Committee*, edited by B. Landmark, New York: Pergamon, 1964, 265–80.

60. Scientific Affairs Division, *NATO and Science*, appendix IV, Projects Involving International Co-operation.

61. Science Committee, Summary Record of Meeting, December 6, 1961, NATO, AC/137-R/11, 13–14. Meeting of radio meteorologists (12/18–21/61), February 17, 1962, NATO, AC/137-D/128.

62. Advisory Panel on Radio-Meteorology, Record of Meeting, April 21, 1965, NATO, AC/137-D/24.

63. A previous version of this section features in Simone Turchetti, "Sword, Shield and Buoys: A History of the NATO Subcommittee on Oceanographic Research, 1959–1973," *Centaurus* 54, no. 3 (2012), 205–31.

64. Samuel Robinson, "Between the Devil and the Deep Blue Sea: Ocean Science and the British Cold War State," PhD diss., University of Manchester, 2015, 172 [to be published as S. Robinson, *Ocean Science and the Cold War State*, New York: Palgrave, 2018].

65. Powaski, *Entangling Alliance*, 66.

66. H. P. Robertson to Howard A. Steadman, December 27, 1960, ROB, box 15, folder 7. On Polaris, see also McDougall, *The Heavens and the Earth*, 326–28.

67. Directorate of Scientific Intelligence, *Oceanography and Defence in the USSR, 1956–58* (London: Ministry of Defence, 1959); copy in the National Archives, Kew Gardens, London, UK [TNA from here on], DEFE 44/27.

68. Norman Friedman, *Seapower as Strategy: Navies and National Interests*, Annapolis, MD: Naval Institute Press, 2001.

69. Camprubí and Robinson, "Gateway to Oceanic Circulation," 437.

70. Kevles, "Cold War, Hot Physics," 245. Robinson, "Between the Devil and the Deep Blue Sea," 140–45; Camprubí and Robinson, "Gateway to Oceanic Circulation," 447–51.

71. Gary Weir, *An Ocean in Common: American Naval Officers, Scientists and the Ocean Environment*, College Station: Texas A&M University Press, 2001, chapters 15 and 16.

72. Ross, "Twenty Years of Research"; Peter Ranelli, "A Little History of the NATO Undersea Research Centre," *Oceanography* 21, no. 2 (2008), 16–20; Tom D. Allan, "Memories from the Sixties," *Oceanography* 21, no. 2 (2008), 21–23.

73. It conveys that secrecy ought to apply to all the countries who are members of the alliance. Kenneth M. Gentry, "Statement of Future Trends in Antisubmarine Warfare," July 17, 1958, Cosmic Top Secret, NATO, AC/137-D/20.

74. "Western military interest in the northern region was in part dictated by the need for denying Soviet forces access to the Atlantic Ocean." Tamnes, "Strategic Importance," 257–74, on 260.

75. Gentry, "Statement of Future Trends," 4.

76. Ibid., 5. On the strategy, see also W. Harriett Critchley, "Polar Deployment of Soviet Submarines," *International Journal* 39 (1984): 828–65.

77. Norman Ramsey, "The Working Group on preparations for a NATO Deep-Sea Oceanographic Expedition," 1958, George Deacon Papers, UK National Oceanography Library, Southampton, UK [DEA from here on], box 87; and Defence Science, April 22, 1959, Confidential, NATO, AC/137-WP/2.

78. Hamblin, *Oceanographers and the Cold War*, chapter 1. See also Torben Wolff, "The Birth and First Years of the Scientific Committee on Oceanic Research," SCOR History Report 1 (2010), 1–96.

79. Wolff, "The Birth and First Years of the Scientific Committee on Oceanic Research," 38.

80. Peder Roberts, "Intelligence and Internationalism: The Cold War Career of Anton Bruun," *Centaurus* 55, no. 3 (2013): 243–63.

81. Weir, *Ocean in Common*, 273.

82. Robinson, "Between the Devil and the Deep Blue Sea," chapter 2. See also Hamblin, *Oceanographers and the Cold War*, chapter 1.

83. On this, see Chandra Mukerji, *A Fragile Power: Scientists and the State*, Princeton, NJ: Princeton University Press, 1989, 43–44.

84. Peder Roberts, "A Frozen Field of Dreams: Science, Strategy, and the Antarctic in Norway, Sweden and the British Empire, 1912–1952," PhD diss., Stanford University, 2010, 175 and 265; see also Eric L. Mills, "From *Discovery* to Discovery: The Hydrology of the Southern Ocean, 1885–1937," *Archives of Natural History* 32 (2005), 246–64.

85. Hamblin, *Oceanographers and the Cold War*, 65. Not unlike Deacon, Lacombe had been employed in sonar-related research during World War II. See Jean Bourgoin, "Henri Lacombe (33), 1913–2000. In Memoriam," *La Jaune et la Rouge* 562 (2001): 51–52.

86. Norman Ramsey, "Recommendations of Ad Hoc Meeting of Experts on Oceanography," March 3, 1959, NATO, AC/137-D/37.

87. Sonar detection and tracking depended on the understanding of the physical properties of the seas. G. Deacon, Speech at the ASW meeting, April 22, 1959, in DEA, box 87.

88. H. A. Sargeaunt to G. Deacon, April 27, 1959, DEA, box 87.

89. Science Committee, Summary Record of Meeting, October 25, 1960, Restricted, NATO, AC/137-R/8, 35.

90. Ramsey, "Recommendations of Ad Hoc Meeting."

91. $227,500 in five years (minus costs for surveys). Torben Wolff, "The Birth and First Years of the Scientific Committee on Oceanic Research," *SCOR History Report* 1 (2010): 38.

92. Ross, "Twenty Years of Research," 19–22.

93. Subcommittee on Oceanographic Research: Progress Report, February 18, 1964, NATO, AC/137-D/208. H. Mosby, "Proposal on Current Measurements in the Faroe-Shetland Channel and in the Straits of Gibraltar," December 31,

1959, Håkon Mosby Papers, University of Bergen, Norway [MOS from here on], 5.

94. Tamnes, "Strategic Importance," 264; Rankin, *After the Map*, 241–42.

95. Camprubí and Robinson, "Gateway to Oceanic Circulation," 457.

96. Ibid., 453. See also Sam Robinson, "Stormy Seas," in *Surveillance Imperative*, edited by Turchetti and Roberts, 105–24.

97. Odd Dahl, "The Capability of the Aanderaa Recording and Telemetering Instrument," *Progress in Oceanography* 5 (1960), 103–6. See also Knut Samset, *Beforehand and Long Thereafter: A Look-Back on the Concept of Some Historical Projects*, Trondheim, Norway: Ex Ante, 2012, 75–77. On Dahl, see also John Krige and Arturo Russo, *A History of the European Space Agency, 1958–1987*, Noordwijk, Neth.: ESA, 44.

98. Robinson, "Stormy Seas," 115–16.

99. "Essentials of the Project Supported by the NATO Research Grant. Turkey, 12 July 1962," DEA, box 87.

100. U. Stefánsonn to G. Deacon, October 6, 1959, DEA, box 87.

101. Naval Steering Group, Summary Record of Meeting, 11 July 1961, Confidential, NATO, AC/141- R/6.

102. Hamblin, *Oceanography and the Cold War*, 187–88.

103. On Nierenberg see also Naomi Oreskes, Erik M Conway, and Matthew Shindell, "From Chicken Little to Dr. Pangloss: William Nierenberg, Global Warming, and the Social Destruction of Scientific Knowledge," *Historical Studies in the Natural Sciences* 38, no. 1 (2008): 109–52.

104. *Activities of the NATO Subcommittee on Oceanographic Research, 1959–1964*, Bergen, Norway: January 1965 (NATO Subcommittee on Oceanographic Research, Technical Report No. 20).

CHAPTER 3

1. Edward Teller, "Nuclear Explosions," in von Kármán Committee [VKC from here on], *Long-Term Studies for the NATO, Third Exercise, Environmental Warfare*, November 1962, Secret, NATO, VKS-EX3- PH1/GP2, 54.

2. Sutcliffe had also edited the proceedings of AGARD's 1956 *Polar Atmosphere Symposium*.

3. As Clark Miller has shown, the WMO was one of the pillars of international relations too, through the provision of expertise and the coordination of national programs under the UN aegis. Clark A. Miller, "Scientific Internationalism in American Foreign Policy: The Case of Meteorology, 1947–1958," in *Changing the Atmosphere: Expert Knowledge and Environmental Governance*, edited by Miller and Paul N. Edwards, 167–218.

4. John Stanley Sawyer, "Reginald Cockcroft Sutcliffe, 16 November 1904–28 May 1991," *Biographical Memoirs of Fellows of the Royal Society* 38 (1992): 347–58, on 354.

5. Ibid. See also Frederik Nebeker, *Calculating the Weather: Meteorology in the*

20th Century, San Diego: Academic Press, 1995; Peter Lynch, "The ENIAC Forecasts: A Re-creation," *Bulletin of the American Meteorological Society* 89, no. 1 (2008): 45–55, on 46–48.

6. On the circumstances and advancement of meteorology, see Kristine Harper, *Weather by the Numbers: The Genesis of Modern Meteorology*, Cambridge, MA: MIT Press, 2008.

7. Miller, "Scientific Internationalism," 167–218, on 167.

8. See chapter 1.

9. Report of Ad Hoc Advisory Group on Meteorology, December 16, 1959, NATO, AC/137-D/57.

10. Report of Ad Hoc Advisory Group on Meteorology, July 7, 1960, NATO, AC/137-D/67.

11. Ibid.

12. DRD, Summary Record of Meeting, November 10, 1960, NATO, AC/137(DR) R/3, 16.

13. The Armand Study Group on increasing the effectiveness of Western Science (1960) in NATO Scientific Affairs Division, *Facts about the Activities of the Science Committee*, Appendix X, 136–39. See also Krige, *American Hegemony*, 210–14.

14. Science Committee, Summary Record of Meeting, September 10, 1960, NATO, AC/137-R/8, 25. Since this is a summary record, it is unclear who made the remark exactly. In the document, the position of personnel at the Scientific Affairs division was added: "the aim should not be to give money to the meteorologists and so on, that would otherwise be given to the chemists and other scientists [. . .] but rather to increase the total effort in the sciences," ibid., 30.

15. Science Committee, Summary Record of Meeting, September 10, 1960, NATO, AC/137-R/8, 36.

16. Franco Foresta Martin and Geppi Calcara, *Per una Storia della Geofisica Italiana*, Milan: Springer, 2010, 145.

17. Giorgio Fea and Adriano Gazzola, "Fenomeni troposferici di trasporto e diffusione a grande scala posti in evidenza da traccianti radioattivi," *Annali di Geofisica* 15, no. 1 (1962): 15–26.

18. The director of Genoa's Geophysical and Geodetic Institute, Mario Bossolasco, had originally advocated the study of cyclones. Hans Volkert, "The International Conferences on Alpine Meteorology: Characteristics and Trends from a 57-Year-Series of Scientific Communication," *Meteorology and Atmospheric Physics* 103 (2009): 5–12.

19. "Discussion, 3 July 1956. Chairman: F. Kenneth Hare," in Sutcliffe, ed., *Polar Atmosphere Symposium*, part 1, *Meteorology Section*, 93–97.

20. Antonio Speranza, "The Formation of Baric Depression near the Alps," *Annals of Geophysics* 28 (1975): 177–217, on 188.

21. NATO Scientific Affairs Division, *NATO and Science: An Account of the Activities of the NATO Science Committee, 1958–1972*, Brussels: NATO, 1973, 112.

22. Trachtenberg, *Constructed Peace*, 287–90. See also Edward Drea, "The Mc-Namara Era" in *History of NATO*, edited by Schmidt, 3: 183–95.

23. In 1970, the UK Minister of Defence at the time, Denis Healey, recalled that the British government had played a leading part in changing NATO doctrine, ruling out the existing "automatic tripwire strategy." Healey, "NATO Use of Tactical Nuclear Weapons," April 6, 1970; copy in ZUC, SZ/2/NATO.

24. S. Zuckerman, "NATO M.R.B.M.s," July 5, 1960, Note for the Minister, D1, personal file marked M/14 "NATO MRBM's," Top Secret, ZUC, SZ/2/NATO. Norstad had announced on that occasion that by 1963 he intended to deploy three hundred Polaris missiles in West Germany. On Norstad's criticism, see Trachtenberg, *Constructed Peace*, 289–90.

25. Peter Goodchild, *Edward Teller: The Real Dr. Strangelove*, Cambridge, MA: Harvard University Press, 2004, 284–87.

26. Peyton, *Solly Zuckerman*, 150.

27. Dietl, "In Defence of the West," 361.

28. L. J. Sabatini to S. Zuckerman, July 6, 1960, ZUC, SZ/2/NATO.

29. MoD Meeting with the Chiefs of Staff, Top Secret—Specially Restricted Circulation, September 19, 196,0, ZUC, SZ/2/NATO.

30. S. Zuckerman to Paul Mason, October 18, 1960, Top Secret, ZUC, SZ/2/NATO.

31. In 1954, Rabi himself, as GAC chairman, had proposed to prioritize disarmament negotiations. Kevles, "Cold War, Hot Physics," 260.

32. Jerry Wiesner to S. Zuckerman, May 31, 1962, Secret, in ZUC, SZ/CSA/133.

33. On the multilateral force, see D. Schwartz, *NATO's Nuclear Dilemmas*, 82–135.

34. Dietl, "In Defence of the West," 370.

35. Ibid., 119. But the force was to be assigned to the NATO SACEUR. See also Richard Moore, *Nuclear Illusion, Nuclear Reality: Britain, the United States and Nuclear Weapons*, New York: Palgrave, 2010.

36. Powaski, *Entangling Alliance*, 65–75.

37. Annex to DRDC, Long-Term Scientific Studies for NATO, Secret, September 14, 1964, NATO, AC/243-D/2.

38. VKC, *Long-Term Studies for the NATO, Interim Report*, June 1961, Secret, NATO, VKS-Interim, iii.

39. Ibid., 5–6.

40. Ibid., 8. On weather manipulation, see James Rodger Fleming, *Fixing the Sky*.

41. VKC, *Long-Term Studies for the Standing Group, NATO: Working Groups 3 and 4*, June 1961, Secret, NATO, VKS-EX2-GP3&4, 15.

42. Ibid., 16.

43. Jacob Darwin Hamblin, *Arming Mother Nature: The Birth of Catastrophic Environmentalism*, Oxford: Oxford University Press, 2013, 139. This would change only in 1977 when the signing of an Environmental Modification Convention (ENMOD) in Geneva by the two superpowers and twenty other countries prohibited the use of environmental warfare. On the emerging knowledge about the effects of fallout, see Rachel Rothschild, "Environmental Awareness

in the Atomic Age: Radioecologists and Nuclear Technology," *Historical Studies in the Physical Sciences* 43, no. 4 (2013): 492–530.

44. VKC, *Long-Term Studies for the Standing Group: Working Groups 3 and 4*, 17.

45. John von Neumann, "Can We Survive Technology?" *Fortune*, June 1955, 106–8.

46. VKC, *Long-Term Studies for the Standing Group: Working Groups 3 and 4*, 95–110. On this, see Zierler, *Invention of Ecocide*.

47. Bruce Lambert, "A. P. Gagge, 85, Biophysicist and Researcher," *New York Times*, February 19, 1993.

48. Edward Teller, "Nuclear Explosions," 54.

49. See on this Gilbert King, "Going Nuclear over the Pacific," *Smithsonian Magazine*, August 15, 2012, https://www.smithsonianmag.com/history/going
-nuclear-over-the-pacific-24428997/.

50. VKC, *Long-Term Studies for the NATO, Third Exercise, Environmental Warfare*, November 1962, Secret, NATO, VKS-EX3- PH1/GP2, 9.

51. Ibid., 16.

52. Fleming, *Fixing the Sky*, chapters 6 and 7. See also Fleming, "The Climate Engineers," *Wilson Quarterly*, Spring 2007, 46–60; Kristine Harper, "Climate Control: United States Weather Modification in the Cold War and Beyond," *Endeavour* 32, no. 1 (2008): 20–26; Harper, "Cold War Atmospheric Sciences in the United States: From Modeling to Control," in *Cold War Science and The Transatlantic Circulation of Knowledge*, edited by Van Dongen, Hoeneveld, and Streefland, 217–43.

53. VKC, *Long-Term Studies for the NATO, Third Exercise*, 39.

54. Ibid., 42.

55. VKC, *Long-Term Studies for the NATO, Supplementary Report*, September 1963, Secret, NATO, VKS-SR, ix.

56. Ibid., 23.

57. Record of a Meeting of the Ad Hoc Advisory Group on Meteorology, February 9, 1963, NATO, AC/137-D/167, 3.

58. "Proposal for an International Institute of Science and Technology," in NATO Scientific Affairs Division, *NATO and Science*, 140–42. See also Krige, *American Hegemony*, 214–17.

59. On this, see Ronald E. Doel and Kristine C. Harper, "Prometheus Unleashed: Science as a Diplomatic Weapon in the Lyndon B. Johnson Administration," in *Global Power Knowledge*, edited by Krige and Barth, 66–86, on 82.

60. Edward Teller, "The Use of Ships and Aircrafts for Meteorological Observations," September 10, 1963, Annex I to Record of a Meeting of the Advisory Group on Meteorology, April 20, 1964, NATO, AC/137-D/210, 9–10.

61. Science Committee, Summary Record of Meeting, December 12, 1963, Restricted, NATO, AC/137-R/17, 32–33.

62. Meeting of the Advisory Group on Meteorology, January 27–28, 1964, NATO, AC/137-D/210, 3.

63. Ibid., 4.

64. Science Committee, Summary Record of Meeting, August 6, 1964, Restricted, NATO, AC/137-R/19, 18–19.
65. Science Committee, Summary Record of Meeting, March 8, 1965, NATO, AC/137-R/21, 11–12.
66. Ibid., 13–14.
67. As shown in Kevles, "Cold War, Hot Physics, 245–47.
68. H. P. Robertson to W. Weaver, June 16, 1960, ROB, box 15, folder 9.
69. Von Kármán still hoped that the NAC would allow the advisory group to play a more prominent role in the shaping of NATO's science policy. Van der Bliek, ed., *AGARD*, 3–22.
70. W. Nierenberg to I. I. Rabi, November 20, 1961, RABI, box 38, folder 10. The project for an IIST modeled on MIT never really took off. See on this Krige, *American Hegemony*, 208–17.
71. W. Nierenberg to I. I. Rabi, March 1, 1962, RABI, box 38, folder 10. The other two individuals mentioned are the NATO Secretary General Dirk U. Stikker and the British nuclear physicist John Cockcroft who was a member of the Armand group.
72. S. Zuckerman to I. I. Rabi, October 22, 1962, RABI, box 38, folder 10.
73. W. Allis to I. I. Rabi, May 2, 1963, RABI, box 38, folder 10.
74. DRD, Provision of Scientific Advice to the NATO Military Authorities, March 29, 1963, Secret, NATO, AC/137(DR)D/3.
75. Letter to the Defence Research Directors from the Exploratory Group, October 17, 1963, Secret, NATO, AC/137(DR)D/4. See also Krige, *American Hegemony*, 205–8, and van der Bliek, ed., *AGARD*, 3–24.
76. The NATO Secretary General, Dirk Stikker, also was dissatisfied with the ways in which the fourth NATO science adviser, William Allis, had administered the research program. Beer, *Integration and Disintegration in NATO*, 227.
77. Letter to the DRDs from the exploratory group, October 17, 1963, Secret, NATO, AC/137(DR)D/4, 28.
78. Ibid.
79. Ibid., 8.
80. See M. Nye, *Michael Polanyi and His Generation*, 213–18; Peyton, *Solly Zuckerman*, 91–92. See also Gary Werskey, *The Visible College: A Collective Biography of British Scientists and Socialists of the 1930s*, London: Free Association, 1988.
81. S. Zuckerman to F. Seitz, February 26, 1960, ZUC, SZ/NATO/3.
82. Krohn, "Solly Zuckerman, 30 May 1904–1 April 1993," 590.
83. Dominique Pestre, "Louis Néel et le Magnétisme à Grenoble: Récit de la Création d'un Empire dans la Province Française, 1940–1965," *Cahiers pour l'Histoire du CNRS* 8 (1990): 52–54.
84. W. Nierenberg, NATO Science Activities, March 4, 1964, RABI, box 38, folder 10, 6.
85. W. Allis to I. I. Rabi, March 10, 1964, in RABI, box 38, folder 10.
86. Howard Simons, "Scientific Activity of NATO is Given Military Emphasis," *Washington Post*, August 1, 1964.

87. Beer, *Integration and Disintegration in NATO*, 208–9.
88. Krige, *American Hegemony*, 96.
89. "Partition of Funds between Member Countries (in US Dollars)," in NATO Scientific Affairs Division, *NATO and Science*, 28.
90. See on this Roberto Cantoni, *Oil Exploration, Diplomacy, and Security in the Early Cold War*, New York: Routledge, 2017, 119–68.
91. Trachtenberg, *Constructed Peace*, 320.
92. Smith, "British Nuclear Weapons and NATO," 1385–6 and 1393–4.
93. The secretary general remarked in his restricted communiqué that documents made available to the Soviets "contained most of the essential elements for the enemy fully to appraise, or obtain confirmation of, NATO's basic defence doctrines and policies." Dirk U. Stikker, "The Pâques Affair," PO/64/118, NATO Secret, March 25, 1964, NATO.
94. G. H. S. Jordan to E. C. Appleyard, May 17, 1965, in TNA, FO 371/189409.
95. Funding of Scientific Programmes, undated [1965] in TNA, FO 371/189409.
96. Science Committee minutes and handwritten notes, February 11, 1965, in TNA, FO 371/189409.
97. Solly Zuckerman, *Scientists and War: The Impact of Science on Military and Civil Affairs*, London: Hamish Hamilton, 1966, 99.
98. Future activities of the committee, February 12, 1965, in TNA, FO 371/189409.
99. H. Brown to J. McLucas, January 12, 1965; copy in RABI, box 38, folder 11.
100. Don Hornig to J. McLucas, December 11, 1964, in RABI, box 38, folder 10.
101. Rabi, memorandum of conversation, undated (August 1964), in RABI, box 39, folder 1.
102. I. I. Rabi to S. Zuckerman, May 7, 1965, ZUC, SZ/CSA/132. Policy studies expert Francis Beer later argued that the "State Department had never been enthusiastic about the McLucas appointment, since it allowed the Department of Defense to dominate the position of Assistant Secretary General for Scientific Affairs." Beer, *Integration and Disintegration in NATO*, 234.
103. S. Zuckerman to I. I. Rabi, January 19, 1965, in RABI, box 38, folder 11.
104. S. Zuckerman to I. I. Rabi, March 23, 1965, and John Gunther Dean, Office of Atlantic Political-Military Affairs to I. I. Rabi, April 14, 1965, in RABI, box 38, folder 11.
105. DRDC, Summary Record of a Meeting, November 18, 1964, Secret, NATO, AC/243-R1, 6.
106. J. McLucas to I. I. Rabi, December 23, 1965, in RABI, box 38, folder 11.
107. S. Zuckerman to I. I. Rabi, September 17, 1981, in RABI, box 9, folder 8.
108. DRDC, Summary Record of a Meeting, November 18, 1964, Secret, NATO, AC/243-R1, 18.
109. Brown to DRDs, September 17, 1965, Restricted, Annex to DRDC, Summary Record of a Meeting, November 10, 1965, Secret, NATO, AC/243-R/3, 37–38.
110. DRDC, Summary Record of a Meeting, November 10, 1965, Secret, NATO, AC/243-R/3, 30.

111. See on this Frederic Bozo, *Two Strategies for Europe. De Gaulle, the United States and the Atlantic Alliance*, Lanham, MD: Rowman and Littlefield, 2002. Powaski, *Entangling Alliance*, 75–76.
112. S. Zuckerman to I. I. Rabi, April 26, 1966, RABI, box 9, folder 8.
113. Zuckerman to Rabi, December 15, 1967, and Rabi to Zuckerman, December 21, 1967, in RABI, box 9, folder 8.

CHAPTER 4

1. Rabi's speech at the tenth anniversary of the foundation of the NATO Science Committee, March 21, 1968, NATO, AC/137-D/330, 24.
2. See Flippen, "Richard Nixon, Russell Train," 613–38, on 615. See also Flippen, *Nixon and the Environment*, Albuquerque: University of New Mexico Press, 2000; Flippen, *Conservative Conservationist: Russell E. Train and the Emergence of American Environmentalism*, Baton Rouge: Louisiana State University Press, 2006.
3. Hamblin, "Environmentalism for the Atlantic Alliance," 54–75. See also Patrick Kyba, "CCMS: The Environmental Connection," *International Journal* 29 (1973): 256–67; Linda Risso, "NATO and the Environment: The Committee on the Challenges of Modern Society," *Contemporary European History* 25, no. 3 (2016), 505–35; and Thorsten Schulz, "Transatlantic Environmental Security in the 1970s? NATO's 'Third Dimension' as an Early Environmental and 'Human Security' Approach," *Historical Social Research* 35 (2010): 309–28. A recent addition is Evanthis Hatzivassiliou, *The NATO Committee on the Challenges of Modern Society, 1969–1975*, New York: Palgrave, 2017.
4. Macekura, *Limits and Growth*, 67. See also S. Macekura, "The Limits of the Global Community: The Nixon Administration and Global Environmental Politics," *Cold War History* 11, no. 4 (2011): 489–518. See also Wolfe, *Competing with the Soviets*, 60–73; McVety, *Enlightened Aid*, chapter 3.
5. The term LDC gained currency from 1964, when it was fist adopted in the context of the UN Conference on Trade and Development (UNCTAD) to identify and classify nations lagging behind in economic growth. See UNCTAD, "The Least Developed Countries: Historical Background," http://www.un.org/events/ldc3/prepcom/history.htm.
6. S. Victor Papacosma, "Greece and NATO: A Nettlesome Relationship," in *History of NATO*, edited by Schmidt, 3: 339–74, on 363.
7. Maintain deterrence and pursue détente, as recalled by Jamie Shea on the occasion of NATO's sixtieth anniversary. See "1967: De Gaulle Pulls France Out of NATO's Integrated Military Structure," https://www.nato.int/cps/en/natohq/opinions_139272.htm.
8. The office was held by Giovanni Polvani (1960–65) and Vincenzo Caglioti (1965–72). On Giacomini, see Giulio Maltese, "Giacomini, Amedeo," *Dizionario Biografico degli Italiani* 54 (2000), www.treccani.it/enciclopedia/amedeo-giacomini_%28Dizionario_Biografico%29/.

9. Rudi Schall, NATO acting science adviser, to J. McLucas, October 24, 1966, in RABI, box 39, folder 1.

10. "Report of Dr. McLucas with Respect to the Scientific Committee Proceedings," October 13, 1966, in RABI, box 41, folder 1.

11. Report by the Exploratory Group of Six, January 30, 1967, Restricted, NATO, AC/137-D/300, 4.

12. Schall was formerly head of the Saint-Louis Institute, a specialized defense laboratory jointly funded by French and West German administrations.

13. A. Martins to I. I. Rabi, March 28, 1967, in RABI, box 39, folder 2.

14. By the mid-1960s, the State Department had also become more disinclined toward traditional policies of development aid targeting African states because of their political side effects, including pushing some toward nonalignment. Macekura, *Limits and Growth*, 67–85.

15. Pollack to Rabi, May 12, 1967, in RABI, box 39, folder 2.

16. Achille Albonetti, "The Technological Gap: Proposals and Documents," in *Spettatore Internazionale* 2, no. 2 (1967): 139–70. See also A. Albonetti, *L'Atomica, l'Italia e l'Europa* (interview by L. Nuti), Rome: Albatros, 2015; Carroll W. Pursell, *Technology in Postwar America: A History*, New York: Columbia University Press, 2007, chapter 7. On the episode, see also Simone Turchetti, "NATO: Cold Warrior or Eco-Warrior?" *Research Europe*, October 28, 2010, 8.

17. Wallace Joyce (acting deputy director of the Office of International Scientific and Technological Affairs) to the undersecretary of state for political affairs, November 7, 1966, National Archives and Records Administration, College Park, MD, USA [NARA from here on], Record Group [RG] 59, box 3074. A collection of US documents on the "technology gap" is now available online at history.state.gov/historicaldocuments/frus1964-68v34.

18. Memorandum from the Interdepartmental Committee on the Technological Gap to President Lyndon Johnson, December 22, 1967, NARA, RG 59, box 3074.

19. "Technological Gap Statistics," undated (1966), in RABI, box 39, folder 6.

20. H. Pollack to I. I. Rabi, May 12, 1967, in RABI, box 39, folder 2.

21. Leopoldo Nuti and Maurizio Cremasco, "Linchpin of the Southern Flank? A General Survey of Italy in NATO, 1949–99," in *History of NATO*, edited by Schmidt, 3: 317–37, on 326–30. See also Alessandro Brogi, "Ike and Italy: The Eisenhower Administration and Italy's 'Neo-Atlanticist' Agenda," *Journal of Cold War Studies* 4, no. 3 (2002): 5–35.

22. Cantoni, *Oil Exploration*, 142.

23. D. Schwartz, *NATO's Nuclear Dilemmas*, 181–86. See also Powaski, *Entangling Alliance*, 75; Russo, *Propaganda and Intelligence*, 108.

24. De Carlo, IBM, to Piore, IBM, August 6, 1966; copy in RABI, box 39, folder 1.

25. R. Schall, "Adaptation of Present Programme to Provide More Effective Aid to Developing NATO Countries," May 24, 1967; copy in RABI, box 41, folder 3.

26. Exploratory Group of Six, Report on Meeting, September 12, 1967, NATO, AC/137–D/315, 9–10.

27. J. McLucas to I. Rabi, November 28, 1967, and I. Rabi to J. McLucas, December 5, 1967, in RABI, box 39, folder 2.

28. There is no clear indication of Brosio's thinking on the subject in his diaries. His opposition is reiterated in M. Brosio to Science Committee members, January 10, 1968, in RABI, box 39, folder 3; see also Beer, *Integration and Disintegration in NATO*, 229.

29. I. Rabi to M. Brosio, January 12, 1968 (draft), and I. Rabi to M. Brosio, January 17, 1968, RABI, box 39, folder 3.

30. Manlio Brosio, *Diari NATO, 1964–1972*, Bologna: Il Mulino, 2011, 485.

31. J. McLucas to J. Killian, January 22, 1968, in RABI, box 39, folder 3.

32. "NATO Science," Limited Official Use, January 6, 1968; copy in RABI, box 41, folder 5, 4.

33. Ibid., 12.

34. Chief Scientific Adviser's Report, Confidential, March 30, 1967, ZUC, SZ/CSA/163.

35. Ibid.

36. John Sheail, "'Torrey Canyon': The Political Dimension," *Journal of Contemporary History* 42, no. 3 (2007): 485–504, on 489. See also Adam Vaughan, "Torrey Canyon Disaster—the UK's Worst-Ever Oil Spill 50 Years On," *Guardian*, March 18, 2017. And, for an interesting perspective "from below," see Timothy Cooper and Anna Green, "The Torrey Canyon Disaster, Everyday Life, and the 'Greening' of Britain," *Environmental History* 22, no. 1 (2017): 101–26.

37. Zuckerman to Donald McLachlan, April 5, 1967, Personal and Confidential, in SZ/CSA/163.

38. "Oil Pollution: No End of a Lesson," *Nature* 219 (1968): 993.

39. S. Zuckerman to PM, September 4, 1968, in ZUC, SZ/CSA/163. See also Political Staff, "Torrey Canyon—PM Blamed for Lack of Action," *Guardian*, August 31, 1968.

40. Environmental Pollution Panel, [US] President's Science Advisory Committee, "Restoring the Quality of Our Environment," White House, November 1965 (excerpts available here: http://dge.stanford.edu/labs/caldeiralab/Caldeira%20downloads/PSAC,%201965,%20Restoring%20the%20Quality%20of%20Our%20Environment.pdf).

41. Robinson, "Between the Devil and the Deep Blue Sea," 258–65.

42. As discussed by the Exploratory Group of Six, Report on Meeting, September 12, 1967, NATO, AC/137-D/315, 6.

43. Speeches commemorating the tenth anniversary of the founding of the NATO Science Committee, March 21, 1968, NATO, AC/137-D/330, 11.

44. Ibid., 17.

45. Ibid., 22–24.

46. Rachel Carson, *Silent Spring*, Cambridge, MA: Riverside, 1962. On Commoner, see Michael Egan, *Barry Commoner and the Science of Survival: The Remaking of American Environmentalism*, Cambridge, MA: MIT Press, 2007.

47. David Farber, *Chicago '68*, Chicago: University of Chicago Press, 1988.
48. Nicholson, *Environmental Revolution*, 276.
49. Macekura, *Limits and Growth,* 95–97.
50. Barbara Ward, *Spaceship Earth*, New York: Columbia University Press, 1966. See also Soraya Boudia, "Observing the Environmental Turn through the Global Environment Monitoring System," in *Surveillance Imperative*, edited by Turchetti and Roberts, 195–212, on 199.
51. Speeches commemorating the tenth anniversary of the foundation of the NATO Science Committee, March 21, 1968, NATO, AC/137-D/330, 25.
52. Finn Aaserud, interview with Gunnar Randers, August 19, 1986, American Institute of Physics, https://www.aip.org/history-programs/niels-bohr-library/oral-histories/32830.
53. Copy of Gunnar Randers's curriculum vitae in RABI, box 42, folder 2. See also Richard G. Hewlett and Jack M. Holl, *Atoms for Peace and War, 1953–1961*, Berkeley, CA: University of California Press, 1989, 619.
54. Macekura, *Limits and Growth*, 105. See also Flippen, *Nixon and the Environment*, chapter 1.
55. On Train, see Flippen, *Conservative Conservationist*.
56. Russell Train, "Speech at the NATO CCMS Committee," April 19, 1971, in Russell Train Papers, US Library of Congress, Washington DC [TRA from here on], box 42, folder 10.
57. US State Department, NATO Committee on the Challenges of Modern Society, December 23, 1970, in TRA, box 42, folder 8. On Nixon's State of the Union address, see Jonathan Movroydis, "The 1971 State of the Union: Nixon's Six Great Goals," January 13, 2016, www.nixonfoundation.org/2016/01/the-1971-state-of-the-union-nixons-six-great-goals/.
58. NATO Information Service, "Address by the President of the United States at the Commemorative NAC Session," April 10, 1969, in the US State Department Foreign Relations of the United States (FRUS), https://history.state.gov/historicaldocuments/frus1969-76v01/d18.
59. For an overview, see Steven R. Weisman, Introduction to *Daniel Patrick Moynihan: A Portrait in Letters of an American Visionary*, New York: Public Affairs, 2010, 3–5.
60. North Atlantic Council, Record of Meeting, May 14, 1969, NATO, C-R(69)22, 5–6.
61. "It shall consider specific problems of the human environment with the deliberate objective of stimulating action by member governments." North Atlantic Council, Memoranda, November 6, 1969, NATO, C-M(69)63.
62. NATO Public Diplomacy Division, *Aspects of NATO—The Challenges of Modern Society*, Brussels: NATO, 1976, 5–7.
63. Hamblin, "Environmentalism for the Atlantic Alliance," 57.
64. Moynihan's "benign neglect" report, on the occasion of the 1969 uprising in the black community, had also exposed the Nixon administration to accusations of racism.

65. The Ford Foundation had, in recent years, begun to sponsor more environ-
    mental organization in the United States as shown in Bosso, *Environment,
    Inc.*, 40–41.

66. Fellowship Awards, Annex II in Status Report of the CCMS Fellowship Pro-
    gramme, September 30, 1975, NATO, AC/274-D/57. William E. Felling, Ford
    Foundation, to R. Train, July 27, 1971, in TRA, box 44, folder 3.

67. CCMS, "Final Report on the Air Pollution Pilot Study," April 26, 1974, NATO,
    AC/274-D/43.

68. In May 1970, Nixon stated that "the oil that fuels our industrial civilization
    can also foul our natural environment." Nixon, speech to US Congress,
    May 20, 1970; copy in TRA, box 44, folder 3.

69. US State Department, NATO Committee on the Challenges of Modern
    Society, December 23, 1970, in TRA, box 42, folder 8.

70. State Department telegram, March 7, 1971, in TRA, box 46, folder 4.

71. US State Department, Limited Official Use, May 13, 1971, NARA, RG 59, box
    2896.

72. Alfred Friendly, "US Fails in Proposal to End Dumping of All Waste in Ocean,"
    Washington Post, April 20, 1971. Russell Train to *Wall Street Journal* director,
    November 30, 1971, in TRA, box 46, folder 6.

73. USNATO, CCMS Plenary Meetings: Press, July 9, 1971, in NATO, RG 59, box
    2897.

74. Hamblin, "Environmentalism for the Atlantic Alliance," 61.

75. Ellsworth, memorandum of conversation, June 30, 1971, in TRA, box 42,
    folder 10.

76. USNATO Telegram 2560, Limited Official Use, June 1971, in NARA, RG 59,
    box 2984. The CCMS conference "Protection of Man and His Environment
    from Meteorological-Hydrological Events" was due to take place on Septem-
    ber 23–24, 1971, at Lake Como.

77. Frederick K. Ling, "Eduard C. Pestel, 1914–1988," *Memorial Tributes: National
    Academy of Engineering* 7 (1994): 182–84.

78. Established in Rome by the Italian scholar and industrialist Aurelio Peccei,
    the club rose to fame in 1972 thanks to its first report, *The Limits of Growth*,
    which forecast a collapse of human society from the projected unsustainable
    development of five global systems (natural resources, population, pollution,
    capital, and agriculture). Club of Rome, *The Limits of Growth*, New York: Uni-
    verse, 1972.

79. Science Committee, Summary Record of Meeting, November 14, 1969, NATO,
    AC/137-R/35, 5–7 and 12–13.

80. Its budget rose from $97,000 in 1969 to $145,000 in 1972. NATO Scientific
    Affairs Division, *NATO and Science*, 28.

81. See "UK/Italy Scientific Co-operation: Anglo-Italian Committee on Science
    and Technology," TNA, FCO 55/497. I am thankful to Giulia Bentivoglio for
    casting light on this bilateral initiative.

82. Powaski, *Entangling Alliance*, 28–42.

83. William Mansfield to Harry Blaney, January 22, 1970, in NARA, RG 59, box 2894.

84. US State Department, NATO Committee on the Challenges of Modern Society, December 23, 1970, in TRA, box 42, folder 8.

85. Speeches commemorating the ten anniversary of the foundation of the NATO Science Committee, March 21, 1968, NATO, AC/137-D/330, 24.

86. NATO Scientific Affairs Division, *NATO and Science*, 27–28.

87. Study of Critical Environmental Problems, ed., *Man's Impact on the Global Environment Assessment and Recommendations for Action*, Cambridge, MA: MIT Press, 1970.

88. Soraya Boudia, "Observing the Environmental Turn," 195–212.

89. Science Committee, Summary Record of Meeting, March 15, 1971, NATO, AC/137-R/39, 10–13.

90. Science Committee, Summary Record of Meeting, December 8, 1971, NATO, AC/137-R/41, 14. Procedures and Regulations for the Subsidiary Bodies of the NATO Science Committee, November 22, 1972, NATO AC/137-D/486, 3–4.

91. Operations research was renamed systems science and a new panel on stress-corrosion cracking also was set up. NATO Scientific Affairs Division, *Scientific Co-operation in NATO: Information on NATO Science Programmes,* Brussels: NATO, 1973, 20.

92. NATO Eco-Sciences Programme, "The History, Advisory Panel on Eco-Sciences (1971–72)," Annex to NATO, ASG.SEA(77)286. Copy in DEA, box 88, folder H8/8.

93. P. D. McTaggart Cowan, "The Canadian North in the Next Hundred Years" *Arctic* 20, no. 4 (1967): 261–62.

94. Advisory Panel on Eco Sciences, Report of a Meeting, 8 September 1971, NATO, AC/137-D/447, 3. On McIntyre see Alison Shaw, "Alasdair McIntyre: Emeritus Professor, Zoologist and Marine Expert," *Herald Scotland,* April 28, 2010, http://www.heraldscotland.com/comment/obituaries/alasdair -mcintyre-emeritus-professor-zoologist-and-marine-expert-1.1023591.

95. Panel on Eco Sciences, Report of a Meeting, September 8, 1971, NATO, AC/137-D/447, 3.

96. Carl H. Oppenheimer, ed., *Environmental Data Management: Conference Proceedings*, New York: Plenum, 1976.

97. See chapter 1. Kenneth Hare went on to pioneer climate change studies and was chairman of Canada's Climate Programme Planning Board. Joan Kenworthy, "Obituary: F. Kenneth Hare," *Weather* 58 (2003): 127–28. On acid rains, see Rachel Rothschild, "Burning Rain: The Long-Range Transboundary Air Pollution Project," in *Toxic Airs: Body, Place, Planet in Historical Perspective*, edited by James Rodger Fleming and Ann Johnson, University of Pittsburgh Press, 2014, 181–207.

98. Science Committee, Summary Record of Meeting, June 26, 1973, NATO, AC/137-R/46, 4–5.

CHAPTER 5

1. Proposal for a Science Committee Conference on the effects of persistent organic substances and heavy metals moving through the ecosystem, September 8, 1972, NATO, Annex II to AC/137-D/473, 7.
2. Kyba, "CCMS: The Environmental Connection," 260.
3. NATO Science Committee, Twentieth Anniversary Commemoration Programme, April 11–13, 1978, Palais d'Egmont, Brussels. Copy in ZUC, SZ/NATO/5.
4. John S. Dryzek, *The Politics of the Earth: Environmental Discourses*, Oxford: Oxford University Press, 1998, 60–63. For prometheanism in environmental history, see, for instance, Stephen Brain, *Song of the Forest: Russian Forestry and Stalinist Environmentalism*, 1905–1953, Pittsburgh, PA: University of Pittsburgh Press, 2011.
5. F. Kenneth Hare "The Environmental Sciences," in NATO Science Committee, twentieth anniversary program; copy in ZUC, SZ/NATO/5.
6. On sustainability, see especially Macekura, *Limits and Growth*, 5.
7. It soon became an international event as other countries followed suit. Macekura, *Limits and Growth*, xi.
8. Manlio Brosio to Permanent NATO Representatives, May 19, 1971, NATO, PO/71/240; copy in TRA, box 42, folder 10. See also Hamblin, "Environmentalism for the Atlantic Alliance," 60.
9. Brosio, *Diari Nato, 1964–72*, Bologna, Italy: Mulino, 774.
10. F. H. Capps, NATO: Support for CCMS Growing, Secret, April 14, 1970, in NARA, RG 59, box 2894.
11. R. Train, memorandum for the president, December 16, 1971, in TRA, box 42, folder 12.
12. "I would prefer no further distribution of this memo until you have read it and we feel it would be ok to have others look at—it's a bit 'frank' on some points." H. Blaney, "Wither CCMS?" Personal and Confidential, March 7, 1972, in TRA, box 42, folder 12.
13. Harry C. Blaney, "NATO's New Challenges to the Problems of Modern Society," *Atlantic Community Quarterly* 11, no. 2 (1973), 246.
14. David D. Dominick, memorandum for Train, September 24, 1970, TRA, box 44, folder 3.
15. A. Denis Clift (White House), memorandum of meeting, December 18, 1970, in TRA, box 46, folder 3.
16. Blaney, memorandum of conversation, April 19, 1971, in TRA, box 46, folder 5.
17. NATIS, "Resolution on the Pollution of the Sea by Oil Spills," December 7, 1970 (http://archives.nato.int/uploads/r/null/1/3/138836/RESOLUTIONS_ON_THE_POLUTION_OF_THE_SEA_BY_OIL_SPILLS_1970-12-07_ENG.pdf).

18. H. Blaney, memorandum for Train, May 4, 1971, in TRA, box 46, folder 5.
19. Details are provided in my article "HMG's Environmentalism: Britain, NATO and the Origins of Environmental Diplomacy," in *Technology, Environment and Modern Britain*, edited by Jon Agar and Jacob Ward, London: UCL Press, 2018, 252–70.
20. Russell E. Train, "A New Approach to International Environmental Cooperation: The NATO Committee on the Challenges of Modern Society," *Kansas Law Review* 22 (1973–74): 167–91, on 175.
21. See MARPOL—25 Years, October 1998, http://www.imo.org/en/Knowledge Centre/ReferencesAndArchives/FocusOnIMO(Archives)/Documents/Focus %20on%20IMO%20-%20MARPOL%20-%2025%20years%20(October %201998).pdf, 2 and 14.
22. F. Hodsoll, memorandum for Train, October 12, 1973, TRA, box 43, folder 5.
23. Charles Lettow, CEQ, to Timothy Atkeson, April 23, 1971, in TRA, box 42, folder 10.
24. This is what the office's experts argued at a 1973 AGARD meeting in London. Turchetti, "HMG's Environmentalism," 264.
25. F. Hodsoll, memorandum for Train, October 12, 1973, TRA, box 43, folder 5. The British attitude to CCMS caused the organization to falter on several occasions. Hamblin, "Environmentalism for the Atlantic Alliance," 62.
26. C. Lettow, CEQ, to T. Atkeson, April 23, 1971; copy in TRA, box 42, folder 10.
27. Tom Winter, CCMS Advanced Automotive Power System, December 3, 1970, TRA, box 44, folder 3.
28. F. Hodsoll, memorandum for Train, November 1, 1972, in TRA, box 43, folder 3.
29. By then Cottrell had replaced Zuckerman as UK chief science adviser. R. Train to A. Cottrell, August 17, 1972, in TRA, box 43, folder 2.
30. CCMS, summary record of a meeting, August 1, 1972, NATO, AC/274-R/9, 13. On Train's role, see Macekura, *Limits and Growth*, 109.
31. Train, "New Approach to International Environmental Cooperation," 167–92, on 189.
32. Kyba, "CCMS: The Environmental Connection," 260.
33. W. H. Mansfield, Garrison Diversion Project, October 18, 1973, in TRA, box 43, folder 5.
34. On the scandal see Keith W. Olson, *Watergate: The Presidential Scandal That Shook America*, Lawrence, KS: University Press of Kansas, 2003.
35. F. Hodsoll, memorandum for Train, November 20, 1973, TRA, box 43, folder 5.
36. State Department, action memorandum, December 3, 1973, NARA, RG 59, box 2897.
37. H. Kissinger to R. Train, February 8, 1974, TRA, box 43, folder 5.
38. Kissinger cited in Train, "New Approach to International Environmental Cooperation," 191.
39. US Department of State, US Representation at the NATO CCMS Spring Plenary, February 8–9, Dusseldorf, Germany; copy in TRA box 43, folder 5, 2.

40. Remarks by the Honorable Russell E. Train at the CCMS Round Table, Brussels, Belgium, October 12, 1976; copy in TRA box 43, folder 5, 2.

41. Science Committee, Summary Record of Meeting, June 1963, Restricted, NATO, AC/137-R/16, 23. On the funding, see "Appendix V: Research Projects Sponsored by the Sub-committee" in NATO Scientific Affairs Division, *NATO and Science*, 104–7.

42. Meeting of the Advisory Panel on Research Grants, June 28, 1963, Restricted, NATO, AC/137-D/182.

43. W. Nierenberg, NATO Science Activities, March 4, 1964, RABI, box 38, folder 10, 6.

44. Science Committee, Summary Record of Meeting, September 1967, AC/137-R/28; copy in DEA, box 87, H8/3.

45. His research focus was initially on the ecosystems of lakes. See Jean-Pierre Descy and François Darchambeau, *Lake Kivu: Limnology and Biogeochemistry of a Tropical Great Lake*, Dordrecht, Neth.: Springer, 2012, 5. On Belgium's activities in marine science, see Roger H. Charlier, "Thirteen Decades of Biological Oceanography in Belgium: Some Highlights (1840s to 1970s)," in *Ocean Science Bridging the Millennia: A Spectrum of Historical Accounts*, edited by the Intergovernmental Oceanographic Commission, Paris: UNESCO, 2004, 369–83.

46. Exploratory Group of Six, Report on Meeting, September 12, 1967, NATO, AC/137-D/315, 9–10.

47. Science Committee, Summary Record of Meeting, June 7, 1971, NATO, AC/137-R/40.

48. Philip Siekevitz and Alfred Sussman, "Oceanography Prostituted by the Military," *Bioscience* 19, no. 7 (1969): 589.

49. Neil Campbell and Edward D. Goldberg, "History and Achievements of the Sub-committee on Oceanographic Research and the Special Programme on Marine Science," NATO Annex to ASG.SEA(77)287, 14; copy in DEA, box 89, H8/9. On Vine, see Naomi Oreskes, "A Context of Motivation: US Navy Oceanographic Research and the Discovery of Sea-Floor Hydrothermal Vents," *Social Studies of Science* 33, no. 5 (2003): 697–742, on 700–10.

50. O. Dahl and H. Mosby, "Guidelines for a Steering Group to Be Responsible for the Preparation of a Final Report to the Science Committee," Report 111, undated, MOS.

51. Campbell and Goldberg, "History and Achievements," 14.

52. H. Mosby, "Tentative Plans and Evaluation for a NATO North Atlantic Platform for Air-Sea Interaction Studies," Report 106, July 21, 1969, MOS.

53. Train, "New Approach to International Environmental Cooperation," 187; CCMS, "Pilot Study on the Pollution of Coastal Waters, 1971," NATO, AC/274-D/8.

54. Lee worked for the UK Ministry of Agriculture, Fisheries, and Food, MAFF, Lowestoft Laboratory. On their disagreements, see Hamblin, *Oceanographers and the Cold War*, 232–35. On Lee's role in the ICES, see Helen M. Rozwa-

dowski, *The Sea Knows No Boundaries: A Century of Marine Science under ICES*, Copenhagen, Denmark: ICES; Seattle: University of Washington Press, 2002.

55. A. Shaler, "Long-Range Programme in Ocean-Atmosphere Interaction," NATO document 6.5.21(70)080/AJS, August 31, 1970, in DEA, box 88, H8/8; Subcommittee on Oceanographic Research, Summary Record of Meeting, January 19, 1971, NATO, AC/137-D/425, 5–7.

56. G. Deacon to A. Capart, June 5, 1970, DEA, box 88, H8/8.

57. A. Shaler, "Role of the Sub-committee on Oceanographic Research, with the Framework of all Activities on the Same Subject," September 2, 1970, DEA, box 88, H8/8.

58. A. Lee, MAFF, to J. D. Bletchly, DSE, November 25, 1970, DEA, box 88, H8/8.

59. Subcommittee on Oceanographic Research, Summary Record of Meeting, April 5, 1971, NATO, AC/137-D/441.

60. J. B. Hersey to G. Deacon, June 15, 1971, box 88, H8/8.

61. P. J. Fallon, DSE, to Miss E. M. Price, MAFF, November 5, 1971, DEA 88, H8/8.

62. Science Committee, Summary Record of Meeting, March 23, 1970, NATO, AC/137-R36, 15.

63. Science Committee, Summary Record of Meeting, December 8, 1971, NATO, AC/137-R/41, 8–10.

64. Ad Hoc Advisory Group on Meteorology, Report on the Fourteenth Meeting, September 26, 1972, NATO, AC/137-D/482.

65. Subcommittee on Oceanographic Research, Report of a Meeting, December 12, 1972, NATO, AC/137-D/489.

66. Subcommittee on Oceanographic Research, Report of a Meeting, November 11, 1971, NATO, AC/137-D/448. Randers's reproach might have been informed by Francis A. Beer's 1969 review of the committee's activities, in which Beer pointed out that "NATO awarded money to individual program directors who then redistributed it without strict supervision by, or strict reporting to, the Organization." Beer, *Integration and Disintegration in NATO*, 215.

67. On Preston's career, see also Jacob Hamblin, *Poison in the Well: Radioactive Waste in the Oceans at the Dawn of the Nuclear Age*, New Brunswick, NJ: Rutgers University Press, 2008, 219–20.

68. A. Preston to G. Deacon, November 10, 1972, DEA, box 89, H8/11.

69. G. Deacon to N. Sanders, DSE, February 27, 1973, and G. Randers to National Delegations, February 28, 1973, ASG.SA(73)070; copies in DEA, box 89, H8/11.

70. G. Deacon to E. W. Thomson, DES, January 2, 1972, in DEA box 89, H8/11.

71. A. Capart to E. Goldberg, May 25, 1973 (cc W. Nierenberg, G. Deacon) in DEA, box 89, H8/11.

72. Suggesting that "Prof. Capart be requested instead to encourage the author of the proposed technical report to seek publication in normal scientific journals whenever possible." Special Programme Panel on Marine Science, Record of a Meeting, March 22, 1974, AC/137-D/541.

73. Karen R. Merrill, *The Oil Crisis of 1973–1974: A Brief History with Documents*, Boston: Bedford/St. Martin's, 2007.
74. Science Committee, "Overview and Summaries of History and Achievements of NATO Science Programmes," September 6, 1978, NATO, AC/137-D/673, 7–8.
75. Science Committee, Summary Record of Meeting, November 29, 1973, NATO, AC/137-R/47, 22.
76. Science Committee, Summary Record of Meeting, July 11, 1974, NATO, AC/137-R/49, 12.
77. Ibid., 7.
78. D. E. Cartwright, "Henry Charnock," *Bibliographical Memoirs of Fellows of the Royal Society* 49 (1999): 36–50.
79. Campbell and Goldberg, "History and Achievements," 7.
80. Ibid., 7.
81. For a comparison, see Oreskes, "Changing the Mission," 141–88.
82. Data set and historical descriptors are available through the British Oceanographic Data Centre: www.bodc.ac.uk/data/documents/series/41183/.
83. As antiknock compounds in fuels for combustion engines (or lead additives).
84. "Cousteau Denuncia i Pericoli Relativi alla *Cavtat*," *Gazzetta del Mezzogiorno*, June 24, 1976.
85. Gian Antonio Danieli, "Commemorazione del Professor Bruno Battaglia," *Atti dell'Istituto Veneto di Scienze, Lettere ed Arti* 170 (2011–12): 111–23.
86. T. D. Allan (executive officer, Marine Sciences Panel) to Clair Patterson, in Clair Cameron Patterson Papers, California Institute of Technology Archive, Pasadena, California [PAT from here on], box 85, folder 10.
87. Clair C. Patterson, "Contaminated and Natural Lead Environments of Man," *Archives of Environmental Health* 11, no. 3 (1965): 344–60. On Patterson, see George R. Tilton, "Clair Cameron Patterson, June 2, 1955–December 5, 1992," *NAS Biographical Memoirs* 74 (1998): 266–87.
88. Giovanni Tiravanti and Gianfranco Boari, "Potential Pollution of a Marine Environment by Lead Alkyls: The *Cavtat* Incident," *Environmental Science and Technology* 13, no. 7 (1979): 849–54, on 851.
89. Special Programme Panel on Marine Sciences, Report of the Sixth Meeting, December 15, 1976, NATO, AC/137-D/623, 4.
90. E. Goldberg to Panel Members, September 7, 1976, PAT, box 85, folder 10.
91. Tiravanti and Boari, "Potential Pollution of a Marine Environment," 853.
92. Alvin Shuster, "Italy, with Eye on Economic Perils, Goes after Sunken Poison Cargo," *New York Times*, March 10, 1977.
93. Special Programme Panel on Marine Sciences, Report of a Meeting, June 5, 1977, NATO, AC/137-D/635, 2.
94. The new panel could do no more than "add a little financial lubrication to international cooperation." J. Sawyer to E. W. Thompson, DSE, January 21, 1972, in DEA, box 89, H8/11.
95. Institute of Oceanographic Sciences, "Wave Measurements during the Joint North Sea Wave Project, RRS John Murray Cruise 9, 5 September–25 Septem-

ber 1973," https://www.bodc.ac.uk/data/information_and_inventories/cruise
_inventory/report/john_murray9_73.pdf. Oceanographers identify the Joint
North Sea Wave Project with the "JONSWAP spectrum," one of the sea state
wave spectra formalized in equations.

96. J. S. Sawyer to DES, March 1, 1974, Restricted. Copy in DEA, box 89, H8/11.
97. J. Gibson, DES to Cabinet Office, FCO, Treasury, UK delegation at NATO,
MoD, MAFF, UK Met Office, and Deacon, May 8, 1974, Restricted. Copy in
DEA, box 89, H8/11.
98. Jamison, *Making of Green Knowledge*, 68–69.
99. Science Committee, Summary Record of Meeting, March 17, 1977, NATO,
AC/137-R/57, 13.

CHAPTER 6

1. Interview with Nils Holme by the author, June 3, 2014.
2. See introduction.
3. "Building weapons, systems, and strategies whose human and machine
components could function as a seamless web, even on the global scales and
in the vastly compressed time frames of superpower nuclear war." Edwards,
Closed World, 1. See also Carroll Pursell, "War in the Age of Intelligent Machines," in *White Heat. People and Technology*, London: BBC, 1994, 144–67.
4. "Proposed Composition and Terms of Reference of the Defence Research
Group," Annex IV to NATO, C-M(66)33 (Revised), 25.
5. Conference of National Armaments Directors, "Status of the Activities of the
Subsidiary Bodies of the DRG," February 14, 1972, NATO, AC/137-D/449.
6. It was eventually disbanded in 1982. NATO Scientific Affairs Division, *NATO
and Science*, 60–64; NATO Scientific Affairs Division, *NATO Science Committee
Yearbook*, Brussels: NATO, 1983, 19.
7. J. Aarons to I. Rabi, January 21, 1965, and the NATO Joint Satellite Group,
Statement of Purpose, undated (1965), in RABI, box 38, folder 11.
8. Mendillo, "Jules Aarons," 7.
9. Aarons, Klobuchar, Allen, and Guidice, "Second Symposium," 164–80, on
175. The research was a contribution to the International Years of the Quiet
Sun, 1964–65, one of the IGY's follow-up initiatives in international scientific cooperation. The International Years of the Quiet Sun took its name
from the series of experiments conducted when the sun reached a minimum
in its irradiation pattern. See UNESCO, *The International Years of the Quiet
Sun*, London: UNESCO, 1964.
10. Aarons, Klobuchar, Allen, and Guidice, "Second Symposium," 164–80.
11. Basu, "In Memoriam: Jules Aarons," 137.
12. Nello Carrara, M. T. De Giorgio, and Pierfranco Pellegrini, "Guide Propagation of HF Radio Waves in the Ionosphere," *Space Science Reviews* 11 (1970):
555–92. US Committee for IQSY-Geophysics Research Board, *United States
Program for the International Years of the Quiet Sun, 1964–65: Interim Progress*

Report, Washington, DC: National Academy of Sciences/National Research Council, March 1965, 16.

13. John P. Mullen, Robin Stafford Allen, and Jules Aarons, "Long Range Propagation Observed on the ORBIS Experiment," *Planetary Space Science* 14 (1966): 155–62.

14. P. F. Checcacci, M. T. de Giorgio, P. Fabeni, G. P. Pazzi, P. F. Pellegrini, A. Ranfagni, and R. Traniello Gradassi, *Ionospheric Propagation Experiments for PAS Satellite*, October 18, 1968, copy of project paper available in Papers of the Consiglio Nazionale delle Ricerche—CNR, Italy, Vincenzo Caglioti (1966–1970) directorship, box 4, folder 68. See also Carrara, De Giorgio, and Pellegrino, "Guide Propagation," 556.

15. NATIS, *Aspects of NATO—Air Defence*, 1, no. 11 (1982), http://archives.nato.int/aspects-of-nato-series-1-n-11-air-defence.

16. NATIS, *Air Defence in a Supersonic Age (NADGE)*, November 1972, http://archives.nato.int/air-defence-in-supersonic-age-nadge.

17. Van der Bliek, ed., *AGARD*, 7–8.

18. Advisory Group on Radio-Meteorology, Summary Record of Meeting, April 21, 1965, NATO, AC/137-D/244, and Summary Record of Meeting, April 15, 1966, NATO, AC/137-D/277. On Brocks, see "Karl Brocks: Dedication," *Boundary-Layer Meteorology* 4 (1973): 3–4.

19. Advisory Group on Radio-Meteorology, Summary Record of Meeting, April 15, 1966, NATO, AC/137-D/418, 4–5.

20. Status of the Activities of the Subsidiary Bodies of the Defence Research Group, February 14, 1972, Confidential, NATO, AC/137-D/449, 7.

21. RSG on Radar Propagation in Ducting Conditions, 1973, NATO, AC/243-D/260.

22. "Status of the Activities of the Subsidiary Bodies of the Defence Research Group," September 16, 1981, Restricted, NATO, AC/137-D/780, 6.

23. Gjessing, *Remote Surveillance*, 1.

24. Ibid., 64.

25. Ibid., 125–30.

26. Arthur P. Cracknell, *Remote Sensing in Meteorology, Oceanography and Hydrology*, Chichester, UK: Ellis Horwood, 1981, 243–47.

27. NATIS, *Aspects of NATO—Air Defence*, 1, no. 11 (1982).

28. Terms of Reference for AC/243(Panel IV/RSG.8) on DRG, Panel on Optics and Infra-red Atmospheric Effects, July 16, 1982, Restricted, Annex II to NATO, AC/243-D/805, 1–4.

29. Ted S. Cress, "Airborne Measurements of Aerosol Size Distribution over Northern Europe, Volume 1. Spring and Fall 1976, Summer 1977," May 29, 1980, USAF Environmental Research Paper n. 702. Air Force Geophysics Lab, Hanscom AFB, MA (a published version of the study appeared in 1981 as Bruce W. Fitch and Ted S. Cress, "Measurements of Aerosol Size Distribution in the Lower Troposphere over Northern Europe," *Journal of Applied Meteorology* 20 (1981), 1119–28.

30. Status of the Activities of the Subsidiary Bodies of the Defence Research Group, September 16, 1981, Restricted, NATO, AC/137-D/780, 24. Also in AC/243(Panel IV/RSG.8) on Atmospheric Optical and IR Effects, Restricted, Annex III to NATO AC/243-D/801. Cress, *Airborne Measurements*.

31. "Status of the Activities of the Subsidiary Bodies," 29.

32. Robert C. Harney, "Military Applications of Coherent Infrared Radar: Physics and Technology of Coherent Infrared Radar," *Physics and Technology of Coherent Infrared Radar (SPIE)* 300 (1981): 2–11.

33. AGARD, *Optical Propagation in the Atmosphere*, Proceedings of a Conference in Lyngby, Denmark, 27–31 October 1975, AGARD Conference Proceedings No. 183, AGARD: Neuilly sur Seine, France, 1976. On Plass's contribution, see Spence Weart, *The Discovery of Global Warming*, Cambridge, MA: Harvard University Press, 2008, 22–25.

34. Robert A. Schiffer, and William B. Rossow, "ISCCP: The First Project of the World Climate Research Programme," *Bulletin of American Meteorological Society* 64 (1983): 779–84. On the uses of radar, see also Hamid Moradkhani, Ruben G. Baird, and Susan A. Wherry, "Assessment of Climate Change Impact on Floodplain and Hydrologic Ecotones," *Journal of Hydrology* 365 (2010): 264–78; Kenneth P. Moran, Brooks E. Martner, M. J. Post, Robert A. Kropfli, David C. Welsh, and Kevin B. Widener, "An Unattended Cloud-Profiling Radar for Use in Climate Research," *Bulletin of the American Meteorological Society* 79, no. 3 (1998): 443–55.

35. "Air-Sea Interaction Panel, Report of a Meeting," May 3, 1977, NATO, AC/137-D/634, 1; NATO Science Committee, *Overview and Summaries of History and Achievements of NATO Science Programmes*, September 6, 1978, NATO, AC/137-D/673, 10.

36. Paul Edwards, *A Vast Machine: Computer Models, Climate Data, and the Politics of Global Warming*, Cambridge, MA: MIT Press, 2010, 251–57.

37. Raymond T. Pollard, Trevor H. Guymer, and Peter K. Taylor, "Summary of the JASIN 1978 Field Experiment," *Philosophical Transactions of the Royal Society of London A* 308 (1983): 221–30.

38. Gjessing, *Remote Surveillance*, iv.

39. CCMS, Summary Record of a Meeting, July 1, 1975, NATO, AC/274-R/17, 29–30. On IMCO, see chapter 4.

40. See Jean-Marie Massin, ed., *Remote Sensing for the Control of Marine Pollution*, New York: Plenum/CCMS, 1984.

41. Annex III to NATO, Air-Sea Interaction Panel, Report of a Meeting, May 3, 1977, AC/137-D/634. See also Omar H. Shemdin, "The West Coast Experiment: An Overview," *EOS* 61, no. 40 (1980): 649–50.

42. See chapter 2.

43. Air-Sea Interaction Panel, Report of a Meeting, May 3, 1977, NATO, AC/137-D/634, 4–5.

44. Caesar Voute, "The Use of Satellites for Verification," in *A Handbook of Verification Procedures*, edited by Frank Barnaby Basingstoke, UK: Palgrave Mac-

millan, 1990, 7–36. France acquired satellite monitoring capabilities only in the late 1980s through the SPOT program. Bhupendra Jasani and Christer Larsson, "Security Implications of Remote Sensing," *Space Policy*, February 1988, 46–59, on 48.

45. ISMA would be "neither feasible nor desirable." Verification "can most effectively be accomplished by parties to agreements themselves." Cited in Walter H. Dorn, *Peace-Keeping Satellites (Peace Research Reviews)* 10, no. 5–6 (1987): 127 and 132.

46. For an overview of possible applications, see G. P. De Loor, "Microwave Measurements over the North Sea," *Boundary-Layer Meteorology* 13 (1978): 119–31; H. Brunsveld van Hulten, "Remote Sensing Program for Oil Detection in the Netherlands," in *Remote Sensing*, edited by Massin, 105–9.

47. Madeleine Godefroy, Gunnar Østrem, and Robin Vaughan, *EARSeL's History: The First Thirty Years*, Hannover, Ger.: EARSeL, 2008, 22.

48. Air-Sea Interaction Panel, Report of a Meeting, May 3, 1977, NATO, AC/137-D/634.

49. Science Committee, Summary Record of Meeting, NATO, November 23, 1977, AC/137-R/59, 22–24.

50. Pollard, Guymer, and Taylor, "Summary of the JASIN 1978 Field Experiment," 221–30.

51. Air-Sea Interaction Panel, Report of a Meeting, January 17, 1978, NATO, AC/137-D/656.

52. Ibid., 4–5.

CHAPTER 7

1. Ex–NATO Secretary General (2009–2014) Anders Fogh Rasmussen, "NATO and Climate Change," *Huffington Post*, 3/18/2010.

2. Jamison, *Making of Green Knowledge*, 93.

3. Science Committee, Summary Record of Meeting, March 17, 1977, NATO, AC/137-R/57, 13.

4. Data from NATO Scientific Affairs Division, *1983 NATO Science Committee Yearbook*, Brussels: NATO, 1983, 30. Data about the 1972 grant budget are from NATO Scientific Affairs Division, *NATO and Science*, 27.

5. Science Committee, *Overview and Summaries of History and Achievements of NATO Science Programmes*, September 6, 1978, NATO, AC/137-D/673, 2.

6. Ibid., 5–6.

7. *Report of the Ad Hoc Group of Expert Advisers on the Future Work of the Science Committee*, January 12, 1979, NATO, AC/137-D/687, 2.

8. Powaski, *Entangling Alliance*, 134. See also Jamie Shea, "1979: The Soviet Union Deploys Its SS20 Missiles and NATO Responds with Its Twin-Track Decision," https://www.nato.int/cps/en/natohq/opinions_139274.htm ?selectedLocale=enl. Kristina Spohr Readman shows that NATO "dual-track decision" drew on a proposal by German chancellor Helmut Schmidt to

modernize its nuclear arsenal, elaborated two years earlier. See Kristina Spohr Readman, "Germany and the Politics of the Neutron Bomb, 1975–1979," *Diplomacy and Statecraft* 21, no. 2 (2010), 259–85, on 279.

9. Leopoldo Nuti, "The Origins of the 1979 Dual Track Decision—A Survey," in *Crisis of Détente*, edited by Nuti, 57–71.

10. On *Able Archer-83* there is an extensive literature. See, for instance, Len Scott, "Intelligence and the Risk of Nuclear War; Able Archer-83 Revisited," *Intelligence and National Security* 26 no. 6 (2011), 759–77 and Arnav Manchanda, "When the Truth Is Stranger than Fiction: The Able Archer Exercise," *Cold War History* 9, no. 1 (2009), 111–33.

11. Science Committee, Summary Record of Meeting, May 2, 1978, NATO, AC/137-R/61, 5.

12. Science Committee, Summary Record of Meeting, August 16, 1978, NATO, AC/137-R/62, 10.

13. Leonard F. Guttridge, *Mutiny: A History of Naval Insurrection*, Annapolis, MD: Naval Institute Press, 1992, 289. See also Alexandros Nafpliotis, *Britain and the Greek Colonels: Accommodating the Junta in the Cold War*, London: I. B. Tauris, 2012.

14. Daniele Ganser, *NATO's Secret Armies*, London: Frank Cass, 2005.

15. *Report of the Ad Hoc Group of Expert Advisers*, 6.

16. Science Committee, Summary Record of Meeting, July 11, 1974, NATO, AC/137-R/49, 5–6.

17. George Contopoulos, *Adventures in Order and Chaos: A Scientific Autobiography*, Dordrecht, Neth.: Kluwer, 2004, 157–58.

18. Shea, "1979."

19. Science Committee, Summary Record of Meeting, April 19, 1979, NATO, AC/137-R/64, 17.

20. Ibid., 4.

21. Science for Stability, Annex to NATO, C-M(79)19, 3–4; NATO Scientific Affairs Division, *Science for Stability Programme: Phase III Post-evaluation 2000*, Brussels: NATO, 2000, 8.

22. Zero (not at all satisfactory) to four (very well). Eight categories (objectives achieved, team development, end-user links, international contacts, dissemination, continued use of results, university courses, uptake of trained people) were listed for each project.

23. NATO Scientific Affairs Division, *Science for Stability Programme: Phase III Post-evaluation 2000*, 65–67.

24. Ibid., 93–94.

25. NATO Scientific Affairs Division, *Science for Stability Programme: Phase II Post-evaluation 1997*, Brussels: NATO, 2000, 49–50.

26. See on this Bocking, *Nature's Experts*, 214–15; see also Leslie R. Alm, *Crossing Borders, Crossing Boundaries: The Role of Scientists in the US Acid Rain Debate*, Westport, CT: Praeger, 2000, 48–52.

27. Rachel Rothschild, "Burning Rain, 181–207; Rothschild, "Acid Wash: How

Cold War Politics Helped Solve a Climate Crisis," *Foreign Affairs* 24 (August 2015), https://www.foreignaffairs.com/articles/2015-08-24/acid-wash.

28. Karen T. Litfin, *Ozone Discourses: Science and Politics in Global Environmental Cooperation*, New York: Columbia University Press, 1994, chapters 3 and 4.

29. F. Kenneth Hare, "Climatic Variation and Variability: Empirical Evidence from Meteorological and Other Sources," in WMO proceedings of the World Climate Conference, Geneva, February 12–32, 1979, 51–81. See also Weart, *Discovery of Global Warming*, 112–13.

30. Science Committee, Summary Record of Meeting, January 25, 1980, NATO, AC/137-R/66, 25.

31. Guthleben, *Histoire du CNRS*, 329–30.

32. NATO Scientific Affairs Division, *1982 NATO Science Committee Yearbook*, Brussels: NATO, 1982, 17–19.

33. Jamison, *Making of Green Knowledge*, 95 and 113–14.

34. Panel on Eco-Sciences, Report of Meeting, January 15, 1979, NATO, AC/137-D/678, 1.

35. Campbell and Goldberg, "History and Achievements."

36. Panel on Marine Sciences, Report of a Meeting, January 23, 1979, NATO, AC/137-D/679.

37. Panel on Marine Sciences, Record of a Meeting, February 5, 1981, NATO, AC/137-D/759.

38. Panel on Air-Sea Interaction, Record of a Meeting, December 19, 1980, NATO, AC/137-D/755, 1.

39. *Proposal for Phasing Out the Special Panel Programme on Air-Sea Interaction*, March 1, 1982, Annex V to AC/137-D/800, 1.

40. R. Chabbal, *Report on Geo Sciences Survey Group*, November 17, 1981, NATO, ASG.SEA(81)327; copy in RABI, box 39, folder 5.

41. Transcript of Paul Merchant's interview with John Houghton, *Oral History of British Science*, British Library, https://sounds.bl.uk/related-content/TRANSCRIPTS/021T-C1379X0045XX-0000A0.pdf, 132–33.

42. Ibid., 134. See also ICSU/WMO, "Scientific Plan for the World Ocean Circulation Experiment," WCRP publication no. 6, Paris: WMO, 1986, www.nodc.noaa.gov/woce/wdiu/wocedocs/sciplan/sciplan.pdf.

43. Transcript of Paul Merchant's interview with John Houghton, 204.

44. Haroun Tazieff, "La Soufrière, Volcanology and Forecasting," *Nature* 269 (1977): 96–97. An overview of the affair is available in the article on the IPG site "À Propos de la Polémique 'Soufrière 1976' . . . ," www.ipgp.fr/~beaudu/soufriere/forum76.html.

45. Responsible for this call was the US volcanologist Alexander McBirney under the pseudonym Derek Bostok. See Derek Bostok, "A Deontological Code for Volcanologists?" *Journal of Volcanology and Geothermal Research* 4 (1978): 1.

46. Guðmundur E. Sigvaldason, "Reply to Editorial," *Journal of Volcanology and Geothermal Research* 4 (1978): I–III.

47. Chabbal, *Report on Geo Sciences Survey Group*, 2.

48. See on this William Glen, ed., *The Mass-Extinction Debates: How Science Works in a Crisis*, Stanford, CA: Stanford University Press, 1994, chapter 1. See also David Sepkoski *Rereading the Fossil Record: The Growth of Paleobiology as an Evolutionary Discipline,* Chicago: University of Chicago Press, 2012, 339–42.

49. Transcript of Paul Merchant's interview with John Houghton, 130–31. Of note in terms of influential publications was a special issue of the journal *Ambio* on the effects of global nuclear exchange, which contained an article by Paul Crutzen and John W. Birks ("The Atmosphere after a Nuclear War: Twilight at Noon," *Ambio* 11, no. 2–3 (1982), 114–25). A recent investigation by *Times* journalist Matt Ridley suggests that the KGB sought to covertly influence the scientific debate to heighten the public scare. See Matt Ridley, "The Russian Role in the Nuclear Winter Theory," February 25, 2018, www .rationaloptimist.com/blog/nuclear-winter/.

50. Chabbal, *Report on Geo Sciences Survey Group*, 3.

51. Since by then he had also become a powerful political figure in France. Lawrence Solomon, "Allegre's Second Thoughts" *Canada National Post*, March 6, 2007.

52. Chabbal, *Report on Geo Sciences Survey Group*, 3–4.

53. Ibid., 3–4. On plate tectonics, see Anthony Hallam, *A Revolution in the Earth Sciences: From Continental Drift to Plate Tectonics*, Oxford: Clarendon, 1973.

54. NATO Scientific Affairs Division, *1983 NATO Science Committee Yearbook*, Brussels: NATO, 1983, 57–58.

55. Weart, *Discovery of Global Warming*, 144–47.

56. NATO Information Service, "A NATO Special Programme on the Science of Global Climate Change," press release 89 (18), May 18, 1989.

57. On the IPCC establishment, see Shardul Agrawala, "Context and Early Origins of the Intergovernmental Panel for Climate Change," *Climatic Change* 39, no. 4 (1998): 605–20; Agrawala, "Structural and Process History of the Intergovernmental Panel for Climate Change," *Climatic Change* 39, no. 4 (1998): 621–42.

58. There is a growing amount of literature on the subject. See Weart, *Discovery of Global Warming*; James Roger Fleming, *Historical Perspectives on Climate Change*, Oxford: Oxford University Press, 1998.

59. H. Nüzhet Dalfes, George Kukla, and Harvey Weiss, eds., *Third Millennium BC Climate Change and Old World Collapse*, Berlin: Springer, 1994.

60. In 1997, he was appointed Minister of Education. Stéphane Foucart, "Claude Allègre, Scientifiquement Incorrect," *Monde*, October 4, 2006.

61. Naomi Oreskes and Erik M. Conway, *Merchants of Doubt: How a Handful or Scientists Obscured the Truth on Issues from Tobacco Smoke to Global Warming*, London: Bloomsbury, 2010, 186–90.

62. Macekura, *Limits and Growth*, 109, 262.

63. Powaski, *Entangling Alliance*, 182.

64. Hamblin, "Environmentalism for the Atlantic Alliance," 70.

65. Bocking, *Nature's Experts*, 222.

66. CCMS, Summary Record of a Meeting, January 19, 1978, NATO, AC/274-R/22, 4–5.

67. Joan D. B. Vandervort, "NATO CCMS 30th Anniversary," Brussels: NATO, 1999, 4.

68. Bosso, *Environment, Inc.*, 101.

69. Wittner, *Toward Nuclear Abolition*, 130–54.

70. NATO CCMS, *The Challenges of Modern Society* (Brussels: Scientific Affairs Division, 1991), http://archives.nato.int/uploads/r/null/1/3/137887/0311 _NATO-CCMS-The_Challenges_of_Modern_Society_1991_ENG.pdf, 25.

71. Ibid., 16.

72. See on this Ulrich Beck, *Risk Society: Towards a New Modernity*, London: Sage, 1992. On the controversy over experts and the Chernobyl disaster, see Bryan Wynne, "Misunderstood Misunderstanding: Social Identities and Public Uptake of Science," *Public Understanding of Science* 1 (1992), 281–304.

73. NATO CCMS, *The Challenges of Modern Society*, 22.

74. Kyba, "CCMS: The Environmental Connection," 260.

75. Vandervort, "NATO CCMS 30th Anniversary," 8–9.

76. Ibid., 10–13.

77. "This time, the Alliance was facing radical, and above all sudden, changes which, with the disappearance of the massive military threat, seemed to deprive the Alliance of the basis of its very existence. [. . .]" Klaus Wittmann, "The Road to NATO's New Strategic Concept," in *History of NATO*, edited by Schmidt, 3: 219–37, on 220.

78. NATO CCMS, "A NATO/CCMS Message to the Montreal Protocol Meeting in Copenhagen, Denmark, on the 17th–25th November 1992," November 16, 1992, NATO, C M(92)60, unclassified. I am grateful to Randi Laura Gebert for providing a copy of this document.

79. Anne Gut and Bruno Vitale, *Contribution au Débat sur l'Uranium Appauvri*, Lausanne: Centre Sanitaire Suisse, 2002.

80. "Sardegna, al Via Processo contro i Veleni del Poligono di Quirra" *Repubblica*, October 29, 2014.

81. Angela Tisbe Ciociola, "Soldati Morti per il Radon sul Venda: Condannato l'ex Direttore Generale," *Corriere della Sera*, November 2, 2017.

CHAPTER 8

1. On opposite ends of the political spectrum, see, for instance, UK Labour leader Jeremy Corbyn (Laura Hughes, Steven Swinford, and Ben Farmer, "Jeremy Corbyn Called NATO to Be Closed Down and Members to 'Give Up, Go Home and Go Away'" *Telegraph* (London), August 16, 2016) and, more recently, US President Donald Trump (Michael Birnbaum, "European Leaders Shocked as Trump Slams NATO and E.U., Raising Fears of Transatlantic Split," *Washington Post*, January 16, 2017).

2. G. F. Kennan, "A Fateful Error," *New York Times*, February 5, 1997, cited in

Zoltan Barany, *The Future of NATO Expansion*, Cambridge, UK: Cambridge University Press, 2003, 16.

3. Jamie Shea, "The Berlin Wall Comes Down and the Soldiers Go Home," https://www.nato.int/cps/en/natohq/opinions_135906.htm.

4. NATO, "Active Engagement, Modern Defence: Strategic Concept for the Defence and Security of the Members of the North Atlantic Treaty Organization adopted by Heads of State and Government in Lisbon, 19 November 2010," www.nato.int/cps/en/natohq/official_texts_68580.htm.

5. Further complicating the understanding of science diplomacy initiatives as always congenial to diplomacy objectives. See AAAS and Royal Society, *New Frontiers*, vi.

6. NATO Unclassified Annex PPC-N(2013)0056-REV1 Enclosure to PO(2013)0230. I am thankful to Randi Laura Gebert of SPS for making available a copy of this document.

7. "NATO-Russia Council," June 16 2017, www.nato.int/cps/ic/natohq/topics_50091.htm, and "NRC Science Committee, NATO-Russia Scientific Cooperation," September 5, 2006, www.nato.int/science/nato-russia/nato-russia.htm.

8. Suzanne Michaelis, "Perspective from NATO," Annex 4 in Environmental and Security initiative (ENVSEC), Report of the Advisory Board Meeting, September 29–30, 2005, http://www.envsec.org/publications/ENVSEC%20Advisory%20Board%20Meeting%20Report%202005.pdf, 22–23.

9. NATO SPS, "Historical Context," May 31, 2011, www.nato.int/science/about_sps/historical.htm.

10. NATO SPS, "Defence and Environment Experts Group," July 19, 2012, http://www.nato.int/cps/en/natolive/88264.htm.

11. NATO SPS, "SILK—Afghanistan: Expanding Internet Connectivity in Afghanistan," May 29, 2013, http://www.nato.int/cps/en/natohq/topics_53359.htm.

12. NATO SPS, "Historical Context."

13. On the mixed philosophy, see chapter 1.

14. NATO Unclassified Annex PPC-N(2013)0056-REV1 Enclosure to PO(2013)0230.

15. Massimo Panizzi, "The Emerging Security Challenges under NATO's New Strategic Concept," November 16, 2011, http://www.nato.int/cps/en/natolive/opinions_81033.htm. See also NATO, "New NATO Division to Deal with Emerging Security Challenges," press release, August 4, 2010, www.nato.int/cps/en/natolive/news_65107.htm.

16. Data for the NATO program appears in "NATO SPS Programme, Annual Report 2013," www.nato.int/nato_static_fl2014/assets/pdf/pdf_topics/20141211_SPS_Annual_Report_2013.pdf, 26; for ESF, see Roland Diehl, see *Multi-Scale Methods for Waves and Transport Processes in Fusion Plasmas: The Legacy of Grigory Pereverzev*, www.ipp.mpg.de/2058252/Diehl.pdf; for ERC, see "ERC Funding Activities, 2007–2013: Key Facts, Patterns and Trends,"

erc.europa.eu/sites/default/files/publication/files/ERC_funding_activities
_2007_2013.pdf. Data on the 2000 budget from NATO Unclassified Annex
PPC-N(2013)0056-REV1 Enclosure to PO(2013)0230.

17. The NERC overall allocation for the financial year 2015–16 was of 40 million
euros ca. (at 2017 Pound Sterling/Euro conversion rates). Data from UK Department of Business, Innovation and Skills, "Science and Research Budget
Allocations for Financial Year 15/16," https://www.gov.uk/government
/uploads/system/uploads/attachment_data/file/278326/bis-14-p200-science
-and-research-budget-allocations-for-2015-to-2016.pdf.

18. Data from: NATO SPS Programme, *Annual Report 2013*, www.nato.int/nato
_static_fl2014/assets/pdf/pdf_topics/20141211_SPS_Annual_Report_2013
.pdf, and NATO SPS Programme, *Annual Report 2014*, http://www.nato.int
/nato_static_fl2014/assets/pdf/pdf_topics/SPS-Annual-Report-2014.pdf.

19. Ibid., 24.

20. "Statement by NATO Foreign Ministers," April 1, 2014, www.nato.int/nrc
-website/en/index.htm.

21. Emerging Security Challenges Division, *NATO SPS Science Day in Kyiv,
Ukraine, 27 May 2016–SPS Cooperation with Ukraine*, Brussels, Belgium: NATO,
2016.

22. SPS Programme, *Annual Report 2014*, 16.

23. For more information, see the webpages of the UNFCCC, http://newsroom
.unfccc.int/.

24. See chapter 5.

25. Data elaborated from UNFCCC table, http://unfccc.int/files/ghg_data
/application/pdf/table.pdf.

26. In particular, French MP Philippe Vitel recalled that "The security of Alliance members is at stake. [. . .] Climate change is increasing the risk of
violent conflict by exacerbating known sources of conflict, like poverty and
economic shocks." See Alex Pashley, "NATO Security 'at Stake' from Climate
Change, Say Lawmakers," October 12, 2015, *Climate Change News*, http://
www.climatechangenews.com/2015/10/12/nato-security-at-stake-from
-climate-change-say-lawmakers/.

27. NATO Parliamentary Assembly, "Resolution 427 on Climate Change and
International Security," https://www.dcaf.ch/sites/default/files/publications
/documents/NPA_resolutions_2015_%D0%95NGLISH.pdf, 25–26. The quote
is in point e in the resolution. I am thankful to the director of NATO's Parliamentary Assembly Science and Technology Committee, Henrik Bliddal, for
pointing me towards this document.

28. Back cover of Oreskes and Conway, *Merchants of Doubt*. See chapter 8, 20–26.

29. See chapter 3.

30. See chapter 6.

31. See chapter 5.

32. Silvio O. Funtowicz and Jerry R. Ravetz, "Science for a Post-normal Age,"
Futures 25, no. 7 (1993): 739–55.

33. Hulme, *Why We Disagree*, 106–8.
34. Chapter 5.
35. Robert Jervis, "Was the Cold War a Security Dilemma?" *Journal of Cold War Studies* 3 (2001): 36–60. See also Charles L. Glaser, "The Security Dilemma Revisited," *World Politics* 50 (1997): 171–201.
36. Chapter 6.
37. Julian Borger, "One False Click . . . ," *Guardian,* January 16, 2016.
38. As a recent standoff between Chinese and US forces off the South China Sea has revealed. Erik Lin-Greenberg, "So China Seized a U.S. Drone Submarine? Welcome to the Future of International Conflict," *Washington Post*, December 23, 2016.
39. "Hacker di Mosca Spiano la Nato e l'Ucraina," *Repubblica*, October 14, 2014.
40. Greg Miller and Adam Entous, "Declassified Report Says Putin 'Ordered' Effort to Undermine Faith in U.S. Election and Help Trump," *Washington Post*, January 6, 2017. More recently, US intelligence agencies have further investigated these covert actions that US President Donald Trump first denied and then confirmed. See David E. Sanger and Matthew Rosenberg, "From the Start, Trump Has Muddied a Clear Message: Putin Interfered," *New York Times*, July 18, 2018.
41. Borger, "One False Click."
42. "NATO Teams Tops Cyber Exercise," April 29, 2015, www.nato.int/cps/en /natohq/news_119085.htm.

Bibliography

AAAS and Royal Society. *New Frontiers in Science Diplomacy: Navigating the Changing Balance of Power*. London: Science Policy Centre, 2010.

Aarons, J., et al. "Atmospheric Phenomena Noted in Simultaneous Observations of 1958 δ2 (Sputnik III)." *Planetary and Space Science* 5 (1961): 169–84.

Aarons, J., J. A. Klobuchar, R. S. Allen, and D. A. Giudice. "Second Symposium on Radio Astronomical and Satellite Studies of the Atmosphere." *IEEE Spectrum*, March 1966, 164–80.

Agar, Jon, "Moon Relay Experiments at Jodrell Bank." In *Beyond the Ionosphere*, edited by Butrica, 19–30.

Agar, Jon. *Science and Spectacle: The Work of Jodrell Bank in Post-war British Culture*, Amsterdam: Harwood Academic, 1998.

Agar, Jon, and Brian Balmer. "British Scientists and the Cold War: The Defence Research Policy Committee and Information Networks, 1947–1963." *Historical Studies in the Physical and Biological Sciences* 28, no. 2 (1998): 209–52.

Agrawala, Shardul. "Context and Early Origins of the Intergovernmental Panel on Climate Change." *Climatic Change* 39, no. 4 (1998): 605–20.

Agrawala, Shardul. "Structural and Process History of the Intergovernmental Panel for Climate Change." *Climatic Change* 39, no. 4 (1998): 621–42.

Albonetti, Achille. "The Technological Gap: Proposals and Documents." *Spettatore Internazionale* 2, no. 2 (1967): 139–70.

Albonetti, Achille. *L'Atomica, l'Italia e l'Europa*. Interview by L. Nuti. Rome: Albatros, 2015.

Allan, Tom D. "Memories from the Sixties." *Oceanography* 21, no. 2 (2008): 21–23.

Alm, Leslie R. *Crossing Borders, Crossing Boundaries: The Role of Scientists in the US Acid Rain Debate*. Westport, CT: Praeger, 2000.

Amadae, Sonjia M. *Rationalizing Capitalist Democracy: The Cold War Origins of Rational Choice Liberalism*. Chicago: University of Chicago Press, 2003.

Anastassiades, M., D. Ilias, G. Moraitis, and P. Giouleas. "Observations Made by the Ionospheric Institute of Athens during the Series of Nuclear Weapon Tests at Novaja Zemlya between 10 September and 4 November 1961." In *Arctic Communications: Proceedings of the Eighth Meeting of the AGARD Ionospheric Research Committee*, edited by B. Landmark, 265–80. New York: Pergamon, 1964.

Aronova, Elena, Baker, Karen and Naomi Oreskes. "Big Science and Big Data in Biology: From the International Geophysical Year through the International Biological Program to the Long-Term Ecological Research Program, 1957–Present," *Historical Studies in the Natural Sciences* 40, no. 2 (2010): 183–224.

Athanassopoulous, Ekavi. *Turkey–Anglo-American Security Interests, 1945–1952: The First Enlargement of NATO*. London: Frank Cass, 1999.

Barany, Zoltan. *The Future of NATO Expansion*. Cambridge, UK: Cambridge University Press, 2003.

Basu, Sunanda. "In Memoriam: Jules Aarons." *IEEE Antennas and Propagation Magazine* 51, no. 1 (2009): 137–38.

Bates, Charles C. "Current Status of Sea Ice Reconnaissance and Forecasting for the American Arctic." In *Polar Atmosphere Symposium*. Part 1, *Meteorology Section*, edited by Sutcliffe, 285–322.

Bates, Charles C., Thomas Frohock Gaskell, and Robert B. Rice. *Geophysics in the Affairs of Man: A Personalized History of Exploration Geophysics and Its Allied Sciences of Seismology and Oceanography*. New York: Pergamon, 1982.

Beck, Ulrich. *Risk Society: Towards a New Modernity*. London: Sage, 1992.

Beer, Francis A. *Integration and Disintegration in NATO: Processes of Alliance Cohesion and Prospects for Atlantic Community*. Columbus: Ohio State University Press, 1969, 204–39.

Blaney, Harry C. "NATO's New Challenges to the Problems of Modern Society." *Atlantic Community Quarterly* 11, no. 2 (1973), 236–47.

Bocking, Stephen. *Nature's Experts: Science, Politics, and the Environment*. New Brunswick, NJ: Rutgers University Press, 2006.

Bosso, Christopher. *Environment, Inc.: From Grassroots to Beltway*. Lawrence: University Press of Kansas, 2005.

Bostok, Derek. "A Deontological Code for Volcanologists?" *Journal of Volcanology and Geothermal Research* 4 (1978): 1.

Boudia, Soraya. "Observing the Environmental Turn through the Global Environment Monitoring System." In *Surveillance Imperative*, edited by Turchetti and Roberts, 195–212.

Bourgoin, Jean. "Henri Lacombe (33), 1913–2000: In Memoriam." *La Jaune et la Rouge* 562 (2001): 51–52.

Bozo, Frederic. *Two Strategies for Europe: De Gaulle, the United States and the Atlantic Alliance*. Lanham, MD: Rowman and Littlefield, 2002.

Brain, Stephen. *Song of the Forest: Russian Forestry and Stalinist Environmentalism, 1905–1953*. Pittsburgh, PA: University of Pittsburgh Press, 2011.

Bright, Cristopher J. *Continental Defense in the Eisenhower Era: Nuclear Antiaircraft Arms and the Cold War*. New York: Palgrave, 2010.

Brogi, Alessandro. "Ike and Italy: The Eisenhower Administration and Italy's 'Neo-Atlanticist' Agenda." *Journal of Cold War Studies* 4, no. 3 (2002): 5–35.

Brosio, Manlio. *Diari NATO, 1964–1972*. Bologna, Italy: Mulino, 2011.

Brunsveld van Hulten, H. W. "Remote Sensing Program for Oil Detection in the Netherlands." In *Remote Sensing for the Control of Marine Pollution*, edited by Massin, 105–9.

Butrica, Andrew J., ed. *Beyond the Ionosphere: Fifty Years of Satellite Communication*. Washington, DC: NASA, 1997.

Camprubí, Lino, and Sam Robinson. "A Gateway to Oceanic Circulation: Submarine Surveillance and the Contested Sovereignty of Gibraltar." *Historical Studies in the Natural Sciences* 46, no. 4 (2016): 429–59.

Cantoni, Roberto. *Oil Exploration, Diplomacy, and Security in the Early Cold War*. New York: Routledge, 2017.

Carrara, Nello, M. T. De Giorgio, and Pierfranco Pellegrini. "Guide Propagation of HF Radio Waves in the Ionosphere." *Space Science Reviews* 11 (1970): 555–92.

Carson, Rachel. *Silent Spring*. Cambridge, MA: Riverside, 1962.

Cartwright, D. E. "Henry Charnock." *Bibliographical Memoirs of Fellows of the Royal Society* 49 (1999): 36–50.

Charlier, Roger H. "Thirteen Decades of Biological Oceanography in Belgium: Some Highlights (1840s to 1970s)." In *Ocean Science Bridging the Millennia: A Spectrum of Historical Accounts*, edited by the Intergovernmental Oceanographic Commission, 369–83. Paris: UNESCO, 2004.

Clavarino, Lodovica. *Scienza e Politica nell'Era Nucleare: La Scelta Pacifista di Edoardo Amaldi*. Rome: Carocci, 2014.

Cloud, John, ed. "Science in the Cold War." Special Issue, *Social Studies of Science* 33, no. 5 (2003).

Club of Rome. *The Limits of Growth*. New York: Universe, 1972.

Cohen-Cole, Jamie. *The Open Mind: Cold War Politics and the Sciences of Human Nature*. Chicago: University of Chicago Press, 2014.

Cohn, Carol. "Sex and Death in the Rational World of Defense Intellectuals." Reprinted in *Women, Science and Technology*, edited by Mary Wyer, Mary Barbercheck, Donna Cookmeyer, Hatice Ozturk, and Marta Wayne, 99–116. London: Routledge, 2001.

Contopoulos, George. *Adventures in Order and Chaos: A Scientific Autobiography*. Dordrecht, Neth.: Kluwer, 2004.

Cooper, Timothy, and Anna Green. "The Torrey Canyon Disaster, Everyday Life, and the 'Greening' of Britain." *Environmental History* 22, no. 1 (2017): 101–26.

Cracknell, Arthur P. *Remote Sensing in Meteorology, Oceanography and Hydrology*. Chichester, UK: Ellis Horwood, 1981.

Crawford, Elisabeth, and Josiane Olff-Nathan, eds. *La Science sous Influence: L'Université de Strasbourg Enjeu des Conflits Franco-Allemands, 1972–1945*. Strasbourg: Nuée Bleue, 2005.

Crawford, Elisabeth, Terry Shinn, and Sverker Sörlin. *Denationalizing Science: The Context of International Scientific Practice*. Dordrecht, Neth.: Kluwer, 1993.

Cress, Ted S. "Airborne Measurements of Aerosol Size Distribution over Northern Europe, Volume 1. Spring and Fall 1976, Summer 1977." May 29, 1980, USAF Environmental Research Paper n. 702. Air Force Geophysics Lab, Hanscom AFB, M. A.Critchley, W. Harriett. "Polar Deployment of Soviet Submarines." *International Journal* 39 (1984): 828–65.

Crutzen, Paul, and John W. Birks. "The Atmosphere after a Nuclear War: Twilight at Noon." *Ambio* 11, no. 2–3 (1982), 114–25.

Dahl, Odd. "The Capability of the Aanderaa Recording and Telemetering Instrument." *Progress in Oceanography* 5 (1960): 103–6.

Dalfes, H. Nüzhet, George Kukla, and Harvey Weiss, eds. *Third Millennium BC Climate Change and Old World Collapse*. Berlin: Springer, 1994.

Danieli, Gian Antonio. "Commemorazione del Professor Bruno Battaglia." *Atti dell'Istituto Veneto di Scienze, Lettere ed Arti* 170 (2011–12): 111–23.

Dawson, Raymond H., and George E Nicholson, "NATO and the SHAPE Technical Center." *International Organization* 21, no. 3 (1967): 565–91.

Day, Michael A. "Oppenheimer and Rabi: American Cold War Physicists as Public Intellectuals." In *The Atomic Bomb and American Society: New Perspectives*, edited by Rosemary B. Mariner and G. Kurt Piehler, 307–28. Knoxville: University of Tennessee Press, 2009.

De Bont, Raf, and Geert Vanpaemel. "The Scientist as Activist: Biology and the Nature Protection Movement, 1900–1950." *Environment and History* 18 (2012): 203–8.

De Loor, G. P. "Microwave Measurements over the North Sea." *Boundary-Layer Meteorology* 13 (1978): 119–31.

De Maria, Michelangelo, and Lucia Orlando. *Italy in Space: In Search of a Strategy, 1957–1975*. Paris: Beauchesne, 2008.

Descy, Jean-Pierre, and François Darchambeau. *Lake Kivu, Limnology and Biogeochemistry of a Tropical Great Lake*. Dordrecht, Neth.: Springer, 2012.

DeWitt, Nicholas. *Soviet Professional Manpower: Training and Supply*. Washington, DC: National Science Foundation, 1955.

Diamond, Louise, and John W. McDonald. *Multi-Track Diplomacy: A Systems Guide and Analysis*. West Hartford, CT: Kumarian, 1996.

Dietl, Ralph. "In Defence of the West: General Lauris Norstad, NATO Nuclear Forces and Transatlantic Relations, 1956–1963." *Diplomacy and Statecraft* 17 (2006): 347–92.

Doel, Ronald E. "Defending the North American Continent: Why the Physical Environmental Sciences Mattered in Cold War Greenland." In *Exploring Greenland, edited by* Doel, Harper, and Heymann, 25–46.

Doel, Ronald E. "Earth Sciences and Geophysics." In *Companion to Science in the 20th Century*, edited by J. Krige and D. Pestre, 391–417. London: Routledge, 2003.

Doel, Ronald E., and Kristine C. Harper. "Prometheus Unleashed: Science as a Diplomatic Weapon in the Lyndon B. Johnson Administration." In *Global Power Knowledge: Science and Technology in International Affairs*, edited by Krige and Barth, 66–86.

Doel, Ronald E., Kristine C. Harper, and Matthias Heymann, eds., *Exploring Greenland: Cold War Science and Technology on Ice*. New York: Palgrave, 2016.

Doel, Ronald E., and Allan A. Needell, "Science, Scientists, and the CIA: Balancing International Ideals, National Needs, and Professional Opportunities." *Intelligence and National Security* 12, no. 1 (1997): 59–81.

Doel, Ronald E., and Naomi Oreskes. "The Physics and Chemistry of the Earth." In *The Cambridge History of Science*, vol. 5, edited by Mary Jo Nye, 538–57. Cambridge: Cambridge University Press, 2012.

Dorn, Walter H. *Peace-Keeping Satellites (Peace Research Reviews)* 10, nos. 5–6 (1987).

Dorsey, Kurkpatrick. *The Dawn of Conservation Diplomacy: U.S.–Canadian Wildlife Protection Treaties in the Progressive Era*. Seattle: University of Washington Press, 1998.

Drea, Edward. "The McNamara Era." In *History of NATO*, edited by Schmidt, 3: 183–95.

Dryzek, John S. *The Politics of the Earth: Environmental Discourses*. Oxford: Oxford University Press, 1998.

Eberle, James. *Life on the Ocean Wave*. Durham, NC: Roundtuit, 2007.

Edgerton, David. *Warfare State: Britain, 1920–1970*. Cambridge, UK: Cambridge University Press, 2006.

Edwards, Paul. *The Closed World: Computers and the Politics of Discourse in Cold War America*. Cambridge, MA: MIT Press, 1996.

Edwards, Paul. *A Vast Machine: Computer Models, Climate Data, and the Politics of Global Warming*. Cambridge, MA: MIT Press, 2010.

Egan, Michael. *Barry Commoner and the Science of Survival: The Remaking of American Environmentalism*. Cambridge, MA: MIT Press, 2007.

Erickson, Paul, Judy L. Klein, Lorraine Daston, Rebecca M. Lemov, and Thomas Sturm. *How Reason Almost Lost Its Mind: The Strange Career of Cold War Rationality*. Chicago: University of Chicago Press, 2013.

Farber, David. *Chicago '68*, Chicago: University of Chicago Press, 1988.

Fea, Giorgio, and Adriano Gazzola. "Fenomeni troposferici di trasporto e diffusione a grande scala posti in evidenza da traccianti radioattivi." *Annali di Geofisica* 15, no. 1 (1962), 15–26.

Fitch, Bruce W., and Ted S. Cress. "Measurements of Aerosol Size Distribution in the Lower Troposphere over Northern Europe." *Journal of Applied Meteorology* 20 (1981), 1119–28.

Fleming, James Rodger. "The Climate Engineers." *Wilson Quarterly*, Spring 2007, 46–60.

Fleming, James Rodger. *Fixing the Sky: The Checkered History of Weather and Climate Control*. New York: Columbia University Press, 2010.

Fleming, James Rodger. *Historical Perspectives on Climate Change*. Oxford: Oxford University Press, 1998.

Fleming, James Rodger. *Inventing the Atmospheric Science: Bjerknes, Rossby, Wexler, and the Foundations of Modern Meteorology*. Cambridge, MA: MIT Press, 2016.

Fleming, James Rodger. "Sverre Petterssen, the Bergen School and the Forecasts for D-Day." *Proceedings of the International Commission on History of Meteorology* 1, no. 1 (2004): 75–83.

Flippen, J. Brooks. *Conservative Conservationist: Russell E. Train and the Emergence of American Environmentalism*. Baton Rouge: Louisiana State University Press, 2006.

Flippen, J. Brooks. "Richard Nixon, Russell Train, and the Birth of Modern American Environmental Diplomacy." *Diplomatic History* 32, no. 4 (2008): 613–38.

Flippen, J. Brooks. *Nixon and the Environment*. Albuquerque: University of New Mexico Press, 2000.

Forman, Paul. "Behind Quantum Electronics: National Security as Basis for Physical Research in the United States, 1940–1960." *Historical Studies in the Physical and Biological Sciences* 18 (1987): 149–229.

Forman, Paul. "Inventing the Maser in Postwar America." *Osiris* 7 (1992): 105–34.

Friedman, Norman. *Seapower as Strategy: Navies and National Interests*. Annapolis, MD: Naval Institute Press, 2001.

Funtowicz, Silvio O., and Jerry R. Ravetz, "Science for a Post-Normal Age." *Futures* 25, no. 7 (1993): 739–55.

Galison, Peter, and Barton Bernstein. "In Any Light: Scientists and the Decision to Build the Superbomb." *Historical Studies in the Physical and Biological Sciences* 19 (1989): 267–347.

Ganser, Daniele. *NATO's Secret Armies*. London: Frank Cass, 2005.

Gieryn, Thomas F. "Boundary Work and the Demarcation of Science from Non-science." *American Sociological Review* 48 (1983): 781–95.

Gieryn, Thomas F. *Cultural Boundaries of Science: Credibility on the Line*. Chicago: University of Chicago Press, 1999.

Gjessing, Dag T. *Remote Surveillance by Electromagnetic Waves for Air-Water-Land*. Ann Arbor, MI: Ann Arbor Science, 1978.

Glaser, Charles L. "The Security Dilemma Revisited." *World Politics* 50 (1997): 171–201.

Glen, William, ed. *The Mass-Extinction Debates: How Science Works in a Crisis*. Stanford, CA: Stanford University Press, 1994.

Godefroy, Madeleine, Gunnar Østrem, and Robin Vaughan. *EARSeL's History: The First Thirty Years*. Hannover, Ger.: EARSeL, 2008.

Goodchild, Peter. *Edward Teller: The Real Dr. Strangelove*. Cambridge, MA: Harvard University Press, 2004.

Gordin, Michael D. *Scientific Babel: How Science Was Done before and after Global English*. Chicago: University of Chicago Press, 2015.

Gorn, Michael H. *The Universal Man: Theodore von Kármán's Life in Aeronautics*. Washington, DC: Smithsonian Institution Press, 1992.

Gottlieb, Robert. *Environmentalism Unbound: Exploring New Pathways for Change*. Cambridge, MA: MIT Press, 2001.

Greenstein, Jesse L. "Howard Percy Robertson, January 27, 1903–August 26, 1961." *NAS Biographical Memoirs*, 51 (1980): 343–426.

Grevsmühl, Sebastian. "Serendipitous Outcomes in Space History: From Space Photography to Environmental Surveillance." In *Surveillance Imperative*, edited by Turchetti and Roberts, 171–94.

Guðmundur E. Sigvaldason. "Reply to Editorial." *Journal of Volcanology and Geothermal Research* 4 (1978): I–III.

Gut, Anne, and Bruno Vitale, *Contribution au Débat sur l'Uranium Appauvri*. Lausanne: Centre Sanitaire Suisse, 2002.

Guthleben, Denis. *Histoire du CNRS de 1939 à Nos Jours*. Paris: Armand Colin, 2009.

Guttridge, Leonard F. *Mutiny: A History of Naval Insurrection*. Annapolis, MD: Naval Institute Press, 1992.

Hallam, Anthony. *A Revolution in the Earth Sciences: From Continental Drift to Plate Tectonics*. Oxford, UK: Clarendon, 1973.

Hamblin, Jacob Darwin. *Arming Mother Nature: The Birth of Catastrophic Environmentalism*. Oxford: Oxford University Press, 2013.

Hamblin, Jacob Darwin. "Environmentalism for the Atlantic Alliance: NATO's Experiment with the 'Challenges of Modern Society.'" *Environmental History* 15 (2010): 54–75.

Hamblin, Jacob Darwin. *Oceanographers and the Cold War: Disciples of Marine Science*. Seattle: University of Washington Press, 2005.

Hamblin, Jacob Darwin. *Poison in the Well: Radioactive Waste in the Oceans at the Dawn of the Nuclear Age*. New Brunswick, NJ: Rutgers University Press, 2008.

Hare, Frederick Kenneth. "Climatic Variation and Variability: Empirical Evidence from Meteorological and Other Sources." In *WMO Proceedings of the World Climate Conference: A Conference of Experts on Climate and Mankind*, Geneva, February 12–23, 1979, 51–81.

Harney, Robert C. "Military Applications of Coherent Infrared Radar: Physics and Technology of Coherent Infrared Radar." *Physics and Technology of Coherent Infrared Radar (SPIE)* 300 (1981): 2–11.

Harper, Kristine. "Climate Control: United States Weather Modification in the Cold War and Beyond." *Endeavour* 32, no. 1 (2008): 20–26.

Harper, Kristine. "Cold War Atmospheric Sciences in the United States: From Modeling to Control." In *Cold War Science and the Transatlantic Circulation of Knowledge*, edited by van Dongen, Hoeneveld, and Streefland, 217–43. Leiden: Brill, 2015.

Harper, Kristine. *Weather by the Numbers: The Genesis of Modern Meteorology*. Cambridge, MA: MIT Press, 2008.

Hatzivassiliou, Evanthis. *The NATO Committee on the Challenges of Modern Society, 1969–1975*. New York: Palgrave, 2017.

Hewlett, Richard G., and Jack M. Holl. *Atoms for Peace and War, 1953–1961*. Berkeley: University of California Press, 1989.

Heymann, Matthias, Henry Nielsen, Kristian Hvidtfelt Nielsen, and Henrik Knudsen. "Small State versus Superpower: Science and Geopolitics in Greenland in the Early Cold War." In *Cold War Science and the Transnational Circulation of Knowledge*, edited by van Dongen, Hoeneveld, and Streefland, 243–71. Leiden: Brill, 2015.

Higuchi, Toshihiro. "Tipping the Scale of Justice: The Fallout Suit of 1958 and the Environmental Legal Dimension of Nuclear Pacifism." *Peace and Change* 38, no. 1 (2013): 33–55.

Hoffman, Andrew J. *How Culture Shapes the Climate Change Debate*. Stanford, CA: Stanford University Press, 2015.

Honick, Morris, and Edd M. Carter. *SHAPE: The New Approach, July 1953–November 1956*. Belgium: SHAPE, 1976.

Hulme, Mike. *Why We Disagree about Climate Change*. Cambridge, UK: Cambridge University Press, 2009.

Jamison, Andrew. *The Making of Green Knowledge: Environmental Politics and Cultural Transformation*. Cambridge, UK: Cambridge University Press, 2001.

Jasani, Bhupendra, and Christer Larsson. "Security Implications of Remote Sensing." *Space Policy* 4, no. 1 (1988): 46–59.

Jensen, Sanne Aagard. "Connecting the Alliance: Communications Infrastructure on the NATO Agenda in the 1950s." In *Historicizing Infrastructure*, edited by Andreas Marklund and Mogens Rüdiger, 183–211. Copenhagen: Aalborg University Press, 2017.

Jervis, Robert. "Was the Cold War a Security Dilemma?" *Journal of Cold War Studies* 3 (2001): 36–60.

Joint Technical Advisory Committee. "Radio Transmission by Ionospheric and Tropospheric Scatter." *IRE Proceedings*, January 1960, 4–44.

Jones, Reginald Victor. *Most Secret War*. London: Penguin, 2009.

Kaiser, David. "The Physics of Spin: Sputnik Politics and American Physicists in the 1950s." *Social Research* 73 (2006): 1225–52.

Kaplan, Lawrence S. *The Long Entanglement: NATO's First Fifty Years*. Westport, CT: Praeger, 1999.

Kaplan, Lawrence S. *The United States and NATO: The Formative Years*. Lexington: University Press of Kentucky, 1984.

Kenworthy, Joan. "Obituary: F. Kenneth Hare." *Weather* 58 (2003): 127–28.

Kevles, Daniel. "Cold War, Hot Physics: Science, Security, and the American State, 1945–56." *Historical Studies in the Physical and Biological Sciences* 20, no. 2 (1990): 239–64.

King, Gilbert. "Going Nuclear over the Pacific." *Smithsonian Magazine*, August 15,

2012, https://www.smithsonianmag.com/history/going-nuclear-over-the
-pacific-24428997/.

Korsmo, Fae L. "The Birth of the International Geophysical Year." *Leading Edge* 10 (2007), 1312–16.

Krige, John. *American Hegemony and the Postwar Reconstruction of Science in Europe.* Cambridge, MA: MIT Press, 2006.

Krige, John. "Atoms for Peace, Scientific Internationalism, and Scientific Intelligence." In *Global Power Knowledge*, edited by Krige and Barth, 161–81.

Krige, John. "Hybrid Knowledge: The Trans-national Co-production of the Gas Centrifuge for Uranium Enrichment in the 1960s." *British Journal for the History of Science* 45 (2012): 337–57.

Krige, John. "I. I. Rabi and the Birth of CERN." *Physics in Perspective* 7 (2005): 150–64.

Krige, John. *Sharing Knowledge, Shaping Europe.* Cambridge, MA: MIT Press, 2016.

Krige, John, and Kai-Henrik Barth, eds. *Global Power Knowledge: Science and Technology in International Affairs.* Chicago: University of Chicago Press, 2006.

Krige, John, and Arturo Russo. *A History of the European Space Agency, 1958–1987.* Noordwijk, Neth.: European Space Agency.

Hermann, Armin, John Krige, Ulrike Mersits, Dominique Pestre, Laura Weiss, and Lanfranco Belloni. *History of CERN.* 3 Vols. Amsterdam: North-Holland, 2000.

Krohn, P. L. "Solly Zuckerman, 30 May 1904–1 April 1993." *Biographical Memoirs of Fellows of the Royal Society* 41 (1995): 575–98.

Kyba, Patrick. "CCMS: The Environmental Connection." *International Journal* 29 (1973): 256–67.

Lalli, Roberto. "'Dirty Work,' but Someone Has to Do It: Howard P. Robertson and the Refereeing Practices of Physical Review in the 1930s." *Notes and Records of the Royal Society of London* 70, no. 2 (2016): 151–74.

Liebowitz, Ruth P. *Chronology: From the Cambridge Field Station to the Air Force Geophysics Laboratory, 1945–1985.* Hanscom, MA: Air Force Geophysics Laboratory, 1985.

Ling, Frederick K. "Eduard C. Pestel, 1914–1988." *Memorial Tributes: National Academy of Engineering* 7 (1994): 182–84.

Litfin, Karen T. *Ozone Discourses: Science and Politics in Global Environmental Cooperation.* New York: Columbia University Press, 1994.

Lord, Kristin M., and Vaughan C. Turekian. "Time for a New Era of Science Diplomacy." *Science* 315 (2007): 769–70.

Lovell, Bernard. *Out of the Zenith: Jodrell Bank, 1957–1970.* New York: Harper and Row, 1973.

Lynch, Peter. "The ENIAC Forecasts: A Re-creation." *Bulletin of the American Meteorological Society* 89, no. 1 (2008): 45–55.

Macekura, Stephen J. "The Limits of the Global Community: The Nixon Administration and Global Environmental Politics." *Cold War History* 11, no. 4 (2011): 489–518.

Macekura, Stephen J. *Of Limits and Growth: The Rise of Global Sustainable Development in the Twentieth Century*. Cambridge, UK: Cambridge University Press, 2015.

Mallard, Grégoire. "L'Europe Puissance Nucléaire, Cet Obscur Objet du Désir." *Critique Internationale* 42 (2009): 141–63.

Malmborg, Mikael af. "Sweden—NATO's Neutral 'Ally'? A Post-revisionist Account." In *History of NATO*, edited by Schmidt, 3: 295–316.

Maloney, Sean M. *Securing Command of the Sea: NATO Naval Planning, 1948–1954*. Annapolis, MD: Naval Institute Press, 1995.

Manchanda, Arnav. "When the Truth Is Stranger than Fiction: The Able Archer Exercise," *Cold War History* 9, no. 1 (2009), 111–33.

Martin, Franco Foresta, and Geppi Calcara. *Per una Storia della Geofisica Italiana*. Milan: Springer, 2010.

Martin-Nielsen, Janet. *Eismitte in the Scientific Imagination: Knowledge and Politics at the Center of Greenland*. New York: Palgrave Macmillan, 2013.

Massin, Jean-Marie, ed. *Remote Sensing for the Control of Marine Pollution*. New York: Plenum/CCMS, 1984.

McDougall, Walter A. . . . *The Heavens and the Earth: A Political History of the Space Age*. New York: Basic Books, 1985.

McGlade, Jacqueline. "NATO Procurement and the Revival of European Defense." In *History of NATO*, edited by Schmidt, 3: 13–28.

McNeill, John R. *Something New under the Sun: An Environmental History of the Twentieth-Century World*. London: Penguin, 2000.

McTaggart-Cowan, Patrick D. "The Canadian North in the Next Hundred Years." *Arctic* 20, no. 4 (1967): 261–62.

McVety, Amanda Kay. *Enlightened Aid: U.S. Development as Foreign Policy in Ethiopia*. Oxford: Oxford University Press, 2012.

Medcalf, Jennifer. *NATO: A Beginner's Guide*. Oxford, UK: One World, 2005.

Mendillo, Michael. "Jules Aarons (1921–2008): Space Weather Pioneer." *Space Weather Quarterly* 6, no. 2 (2009): 6–7.

Merrill, Karen R. *The Oil Crisis of 1973–1974: A Brief History with Documents*. Boston: Bedford/St. Martin's, 2007.

Miller, Clark A., and Paul N. Edwards, eds. *Changing the Atmosphere: Expert Knowledge and Environmental Governance*. Cambridge, MA: MIT Press, 2001.

Miller, Clark A. "Scientific Internationalism in American Foreign Policy: The Case of Meteorology, 1947–1958." In *Changing the Atmosphere: Expert Knowledge and Environmental Governance*, edited by Miller and Edwards, 167–218.

Mills, Eric L. *The Fluid Envelope of Our Planet: How the Study of Ocean Currents Became a Science*. Toronto: University of Toronto Press, 2011.

Mills, Eric L. "From *Discovery* to Discovery: The Hydrology of the Southern Ocean, 1885–1937." *Archives of Natural History* 32 (2005): 246–64.

Moedas, Carlos. "Science Diplomacy in the European Union." *Science and Diplomacy* 5, no. 1 (2016).

Moore, Richard. *Nuclear Illusion, Nuclear Reality: Britain, the United States and Nuclear Weapons*. New York: Palgrave, 2010.

Moradkhani, Hamid, Ruben G. Baird, and Susan A. Wherry. "Assessment of Climate Change Impact on Floodplain and Hydrologic Ecotones." *Journal of Hydrology* 365 (2010): 264–78.

Moran, Kenneth P., Brooks E. Martner, M. J. Post, Robert A. Kropfli, David C. Welsh, and Kevin B. Widener. "An Unattended Cloud-Profiling Radar for Use in Climate Research." *Bulletin of the American Meteorological Society* 79, no. 3 (1998): 443–55.

Mukerji, Chandra. *A Fragile Power: Scientists and the State*. Princeton, NJ: Princeton University Press, 1989.

Mullen, John. P., Robin Stafford Allen, and Jules Aarons. "Long Range Propagation Observed on the ORBIS Experiment." *Planetary Space Science* 14 (1966): 155–62.

Nafpliotis, Alexandros. *Britain and the Greek Colonels: Accommodating the Junta in the Cold War*. London: I. B. Tauris, 2012.

NATO Information Service. *Facts about the North Atlantic Treaty Organization*. Utrecht, Neth.: NATO/Bosch, 1962.

NATO Public Diplomacy Division. *Aspects of NATO — The Challenges of Modern Society*. Brussels: NATO, 1976.

NATO Scientific Affairs Division. *NATO and Science: An Account of the Activities of the NATO Science Committee, 1958–1972*. Brussels: NATO, 1973.

NATO Scientific Affairs Division. *1982 NATO Science Committee Yearbook*. Brussels: NATO, 1982.

NATO Scientific Affairs Division. *1983 NATO Science Committee Yearbook*. Brussels: NATO, 1983.

NATO Scientific Affairs Division. *Science for Stability Programme: Phase II Post-evaluation 1997*. Brussels: NATO, 2000.

NATO Scientific Affairs Division. *Science for Stability Programme: Phase III Post-evaluation 2000*. Brussels: NATO, 2000.

NATO Scientific Affairs Division. *Scientific Co-operation in NATO: Information on NATO Science Programmes*. Brussels: NATO, 1973.

Nebeker, Frederik. *Calculating the Weather: Meteorology in the 20th Century*. San Diego, CA: Academic Press, 1995.

Needell, Allan. *Science, Cold War and the American State: Lloyd V. Berkner and the Balance of Professional Ideals*. Amsterdam: Harwood Academic, 2000.

Nicholson, Max. *The Environmental Revolution: A Guide for the New Masters of the World*. Harmondsworth, UK: Penguin, 1970.

Nilsson, Mikael. "Amber Nine: NATO's Secret Use of a Flight Path over Sweden and the Incorporation of Sweden in NATO's Infrastructure." *Journal of Contemporary History* 44, no. 2 (2009): 287–307.

Nuti, Leopoldo, ed. *The Crisis of Détente in Europe: From Helsinki to Gorbachev, 1975–1985*. London: Routledge, 2008.

Nuti, Leopoldo. "The Origins of the 1979 Dual Track Decision—A Survey." In *Crisis of Détente*, edited by Nuti, 57–71.

Nuti, Leopoldo. *La Sfida Nucleare: La Politica Estera Italiana e le Armi Atomiche, 1945–1991*. Bologna, Italy: Mulino, 2007.

Nuti, Leopoldo, and Maurizio Cremasco. "Linchpin of the Southern Flank? A General Survey of Italy in NATO, 1949–99." In *History of NATO*, edited by Schmidt, 3: 317–37.

Nye, Joseph. *Soft Power: The Means to Success in World Politics*. New York: Public Affairs, 2004.

Nye, Mary Jo. *Michael Polanyi and His Generation: Origins of the Social Construction of Science*. Chicago: University of Chicago Press, 2011.

"Oil Pollution: No End of a Lesson." *Nature* 219 (1968): 993.

Olson, Keith W. *Watergate: The Presidential Scandal That Shook America*. Lawrence: University Press of Kansas, 2003.

Omay, C. Güner. *Prof. Dr. Egbert Adriaan Kreiken: Founder of the Ankara University Observatory and a Volunteer of Education*. Birinci Basım, Turkey: Kasim, 2011.

Oppenheimer, Carl H., ed. *Environmental Data Management: Conference Proceedings*. New York: Plenum, 1976.

Oreskes, Naomi. "Changing the Mission: From Cold War to Climate Change." In *Science and Technology in the Global Cold War*, edited by Oreskes and Krige, 141–88.

Oreskes, Naomi. "A Context of Motivation: US Navy Oceanographic Research and the Discovery of Sea-Floor Hydrothermal Vents." *Social Studies of Science* 33, no. 5 (2003): 697–742.

Oreskes, Naomi. "Science in the Origins of the Cold War." In *Science and Technology in the Global Cold War*, edited by Oreskes and Krige, 11–24.

Oreskes, Naomi, and Erik M. Conway. *Merchants of Doubt: How a Handful or Scientists Obscured the Truth on Issues from Tobacco Smoke to Global Warming*. London: Bloomsbury, 2010.

Oreskes, Naomi, Erik M. Conway, and Matthew Shindell. "From Chicken Little to Dr. Pangloss: William Nierenberg, Global Warming, and the Social Destruction of Scientific Knowledge." *Historical Studies in the Natural Sciences* 38, no. 1 (2008): 109–52.

Oreskes, Naomi, and John Krige, eds. *Science and Technology in the Global Cold War*. Cambridge, MA: MIT Press, 2014.

Paár-Jákli, Gabriella. *Networked Governance and Transatlantic Relations: Building Bridges through Science Diplomacy*. New York: Routledge, 2014.

Papacosma, S. Victor. "Greece and NATO: A Nettlesome Relationship." In *History of NATO*, edited by Schmidt, 3: 339–74.

Patterson, Clair C. "Contaminated and Natural Lead Environments of Man." *Archives of Environmental Health* 11, no. 3 (1965): 344–60.

Pedaliu, Effie G. H. *Britain, Italy and the Origins of the Cold War*. New York: Palgrave, 2003.

Pestre, Dominique. "Louis Néel et le Magnétisme à Grenoble: Récit de la Création d'un Empire dans la Province Française, 1940–1965." *Cahiers pour l'Histoire du CNRS* 8 (1990): 52–54.

Pestre, Dominique. "Studies of the Ionosphere and Forecasts for Radiocommunications: Physicists and Engineers, the Military and National Laboratories in France (and Germany) after 1945." *History and Technology* 13, no. 3 (1997): 183–205.

Petersen, Nikolaj. "Denmark's Fifty Years with NATO." In *History of NATO*, edited by Schmidt, 3: 275–94.

Peyton, John. *Solly Zuckerman: A Scientist Out of the Ordinary*. London: John Murray, 2001.

Pollard, Raymon T., Trevor H. Guymer, and Peter K. Taylor. "Summary of the JASIN 1978 Field Experiment." *Philosophical Transactions of the Royal Society of London* 308 (1983): 221–30.

Powaski, Ronald E. *The Entangling Alliance: The United States and European Security, 1950–1993*. Westport, CT: Greenwood, 1994.

Pursell, Carroll W. *Technology in Postwar America: A History*. New York: Columbia University Press, 2007.

Pursell, Carroll W. "War in the Age of Intelligent Machines." In *White Heat: People and Technology*. London: BBC, 1994, 144–67

Ranelli, Peter. "A Little History of the NATO Undersea Research Centre." *Oceanography* 21, no. 2 (2008): 16–20.

Rankin, William. *After the Map: Cartography, Navigation, and the Transformation of Territory in the Twentieth Century*. Chicago: University of Chicago Press, 2016.

Rau, Erik P. "Technological Systems, Expertise, and Policy Making: The British Origins of Operational Research." In *Technologies of Power: Essays in Honor of Thomas Parker Hughes*, edited by Michael Thad Allen and Gabrielle Hecht, 215–52. Cambridge, MA: MIT Press, 2001.

Readman, Kristina Spohr. "Germany and the Politics of the Neutron Bomb, 1975–1979," *Diplomacy and Statecraft* 21, no. 2 (2010), 259–85.

Richelson, Jeffrey. *Spying on the Bomb: American Nuclear Intelligence from Nazi Germany to Iran and North Korea*. New York: Norton, 2006.

Rigden, John S. *Rabi: Scientist and Citizen*. Cambridge, MA: Harvard University Press, 1987.

Risso, Linda. "NATO and the Environment: The Committee on the Challenges of Modern Society." *Contemporary European History* 25, no. 3 (2016): 505–35.

Roberts, Peder. "A Frozen Field of Dreams: Science, Strategy, and the Antarctic in Norway, Sweden and the British Empire, 1912–1952." PhD diss., Stanford University, 2010.

Roberts, Peder. "Intelligence and Internationalism: The Cold War Career of Anton Bruun." *Centaurus* 55, no. 3 (2013): 243–63.

Roberts, Peder. "Scientists and Sea Ice Surveillance in the Early Cold War." In *Surveillance Imperative*, edited by Turchetti and Roberts, 125–44.

Robertson, Thomas. *The Malthusian Moment: Global Population Growth and the Birth of American Environmentalism*. New Brunswick, NJ: Rutgers University Press, 2012.

Robinson, Samuel. "Between the Devil and the Deep Blue Sea: Ocean Science and the British Cold War State." PhD diss., University of Manchester, 2015.

Robinson, Samuel. "Stormy Seas: Anglo-American Negotiations on Ocean Surveillance." In *Surveillance Imperative*, edited by Turchetti and Roberts, 105–24.

Ross, Donald. *Twenty Years of Research at the SACLANT ASW Research Centre, SACLANTCEN*. Special Report M-93, January 1, La Spezia, Italy: NATO, 1980.

Rothschild, Rachel. "Acid Wash: How Cold War Politics Helped Solve a Climate Crisis." *Foreign Affairs* 24 (August 2015).

Rothschild, Rachel. "Burning Rain: The Long-Range Transboundary Air Pollution Project." In *Toxic Airs: Body, Place, Planet in Historical Perspective*, edited by James Rodger Fleming and Ann Johnson, 181–207. Pittsburgh: University of Pittsburgh Press, 2014.

Rothschild, Rachel. "Environmental Awareness in the Atomic Age: Radioecologists and Nuclear Technology." *Historical Studies in the Physical Sciences* 43, no. 4 (2013): 492–530.

Rozwadowski, Helen M. *The Sea Knows No Boundaries: A Century of Marine Science under ICES*. Copenhagen: ICES; Seattle: University of Washington Press, 2002.

Russo, Linda. *Propaganda and Intelligence in the Cold War: The NATO Information Service*. London: Routledge, 2014.

Samoggia, Franco. *Nello Carrara*. Siena, Italy: Carlo Cambi, 2006.

Samset, Knut. *Beforehand and Long Thereafter: A Look-Back on the Concept of Some Historical Projects*. Trondheim, Norway: Ex Ante, 2012.

Saunier, Pierre Y., and Akira Iriye, eds. *The Palgrave Dictionary of Transnational History*. New York: Palgrave, 2009.

Sawyer, John Stanley. "Reginald Cockcroft Sutcliffe, 16 November 1904–28 May 1991." *Biographical Memoirs of Fellows of the Royal Society* 38 (1992): 347–58.

Schiffer, Robert A., and William B. Rossow. "ISCCP: The First Project of the World Climate Research Programme." *Bulletin of American Meteorological Society* 64 (1983): 779–84.

Schmidt, Gustav, ed. *A History of NATO: The First Fifty Years*. Basingstoke, UK: Palgrave, 2001.

Schulz, Thorsten. "Transatlantic Environmental Security in the 1970s? NATO's 'Third Dimension' as an Early Environmental and 'Human Security' Approach." *Historical Social Research* 35 (2010): 309–28.

Schwartz, David N. *NATO's Nuclear Dilemmas*. Washington, DC: Brookings Institution Press, 1983.

"Science and Foreign Relations: Berkner Report to the U.S. Department of State." *Bulletin of the Atomic Scientists* 6, no. 10 (1950), 293–98.

Scott, Len. "Intelligence and the Risk of Nuclear War; Able Archer-83 Revisited." *Intelligence and National Security* 26, no. 6 (2011), 759–77.

Sebesta, Lorenza. "American Military Aid and European Rearmament." In *NATO: The Founding of the Atlantic Alliance and the Integration of Europe*, edited by Francis H. Heller and John R. Gillingham, 283–310. Basingstoke, UK: Macmillan, 1992.

Sebesta, Lorenza. "Italian Space Policy between Internal Renovation and External Challenges at the Turn of the Decade, 1957–1963." In *Italy in Space: In Search of a Strategy, 1957–1975*, edited by Michelangelo De Maria and Lucia Orlando, 47–76, Paris: Beauchesne, 2008.

Sepkoski, David. *Rereading the Fossil Record: The Growth of Paleobiology as an Evolutionary Discipline*. Chicago: University of Chicago Press, 2012.

Sheail, John. "'Torrey Canyon': The Political Dimension." *Journal of Contemporary History* 42, no. 3 (2007): 485–504.

Shemdin, Omar H. "The West Coast Experiment: An Overview." *EOS* 61, no. 40 (1980): 649–50.

Siekevitz, Philip, and Alfred Sussman. "Oceanography Prostituted by the Military." *Bioscience* 19, no. 7 (1969): 589.

Smith, Frank L. III. "Advancing Science Diplomacy: Indonesia and the US Naval Medical Research Unit." *Social Studies of Science* 44, no. 6 (2014): 825–47.

Smith, Martin A. "Britain Nuclear Weapons and NATO in the Cold War and Beyond." *International Affairs* 87, no. 6 (2011): 1385–99.

Sörlin, Sverker. "Ice Diplomacy and Climate Change: Hans Ahlmann and the Quest for a Nordic Region beyond Borders." In *Science, Geopolitics and Culture in the Polar Region*, edited by Sörlin, 23–55. Farnham, UK: Ashgate, 2013.

Sounders, Frances Stonor. *The Cultural Cold War*. New York: New Press, 1999.

Speranza, Antonio. "The Formation of Baric Depression near the Alps." *Annals of Geophysics* 28 (1975): 177–217.

Stoddart, Kristan. *The Sword and the Shield: Britain, America, NATO, and Nuclear Weapons, 1970–1976*. New York: Palgrave, 2014.

Study of Critical Environmental Problems, ed. *Man's Impact on the Global Environment Assessment and Recommendations for Action*. Cambridge, MA: MIT Press, 1970.

Sullivan, Walter. *Assault on the Unknown: The International Geophysical Year*. New York: McGraw-Hill, 1961.

Sutcliffe, Reginald C., ed. *Polar Atmosphere Symposium. Part 1, Meteorology Section*. New York: Pergamon, 1958.

Tamnes, Rolf. "The Strategic Importance of the High North during the Cold War." In *History of NATO*, edited by Schmidt, 3: 257–74.

Tazieff, Haroun. "La Soufrière, Volcanology and Forecasting." *Nature* 269 (1977): 96–97.

Thomas, William. *Rational Action: The Sciences of Policy in Britain and America, 1940–1960*. Cambridge, MA: MIT Press, 2015.

Tilton, George R. "Clair Cameron Patterson, June 2, 1955–December 5, 1992." *NAS Biographical Memoirs* 74 (1998): 266–87.

Tiravanti, Giovanni, and Gianfranco Boari. "Potential Pollution of a Marine Environment by Lead Alkyls: The *Cavtat* Incident." *Environmental Science and Technology* 13, no. 7 (1979): 849–54.

Trachtenberg, Marc. *A Constructed Peace: The Making of the European Settlement, 1945–1963.* Princeton, NJ: Princeton University Press, 1999.

Train, Russell E. "A New Approach to International Environmental Cooperation: The NATO Committee on the Challenges of Modern Society." *Kansas Law Review* 22 (1973–74): 167–91.

Turchetti, Simone. "'In God We Trust, All Others We Monitor': Seismology, Surveillance and the Test Ban Negotiations." In *Surveillance Imperative*, edited by Turchetti and Roberts, 85–104.

Turchetti, Simone. "A Most Active Customer: How the U.S. Administration Helped the Italian Atomic Energy Project to 'De-develop.'" *Historical Studies in the Natural Sciences* 44, no. 5 (2014): 470–502.

Turchetti, Simone. "NATO: Cold Warrior or Eco-Warrior?" *Research Europe*, October 28, 2010, 8.

Turchetti, Simone. "A Need-to-Know-More Criterion? Science and Information Security at NATO during the Cold War." In *Cold War Science and the Transatlantic Circulation of Knowledge*, edited by J. Van Dongen, F. Hoeneveld, and A. Streefland, 36–58. Leiden: Brill, 2015.

Turchetti, Simone. *The Pontecorvo Affair: A Cold War Defection and Nuclear Physics.* Chicago: University of Chicago Press, 2012.

Turchetti, Simone. "Sword, Shield and Buoys: A History of the NATO Subcommittee on Oceanographic Research, 1959–1973." *Centaurus* 54, no. 3 (2012): 205–31.

Turchetti, Simone, Néstor Herran, and Soraya Boudia. "Introduction: Have We Ever Been 'Transnational'? Towards a History of Science across and beyond Borders." *British Journal for the History of Science* 45 (2012): 319–36.

Turchetti, Simone, Simon Naylor, Katrina Dean, and Martin Siegert. "On Thick Ice: Scientific Internationalism and Antarctic Affairs, 1957–1980." *History and Technology* 24, no. 4 (2008): 351–76.

Turchetti, Simone, and Peder Roberts, eds. *The Surveillance Imperative: Geosciences during the Cold War and Beyond.* New York: Palgrave, 2014.

Turekian, Vaughan C., and Norman P. Neureiter. "Science and Diplomacy: The Past as Prologue." *Science and Diplomacy* 1 (2012).

UNESCO. *The International Years of the Quiet Sun.* London: UNESCO, 1964.

US Committee for IQSY-Geophysics Research Board. *United States Program for the International Years of the Quiet Sun, 1964–65: Interim Progress Report.* Washington, DC: National Academy of Sciences/National Research Council, March 1965, 16.

Vance, Tiffany C., and Ronald E. Doel. "Graphical Methods and Cold War Scientific Practice: The Stommel Diagram's Intriguing Journey from the Physical to the Biological Environmental Sciences." *Historical Studies in the Natural Sciences* 40 (2010): 1–47.

Van der Bliek, Jan, ed. *AGARD: The History, 1952–1997*. Illford, UK: RTO/STS Communications, 1999.

Vandervort, Joan D. B. "NATO CCMS 30th Anniversary." Brussels: NATO, 1999.

Van Dongen, Jeroen, and Friso Hoeneveld, "*Quid pro Quo*: Dutch Defense Research during the Early Cold War." In *Cold War Science and the Transnational Circulation of Knowledge*, edited by van Dongen, Hoeneveld, and Streefland, 101–21. Leiden: Brill, 2015.

Volkert, Hans. "The International Conferences on Alpine Meteorology: Characteristics and Trends from a 57-Year-Series of Scientific Communication." *Meteorology and Atmospheric Physics* 103 (2009): 5–12.

Von Neumann, John. "Can We Survive Technology?" *Fortune*, June 1955, 106–8. Reprinted in *The Fabulous Future: America in 1980, edited by* David Sarnoff, 33–48. New York: Dutton, 1956.

Voute, Caesar. "The Use of Satellites for Verification." In *Handbook of Verification Procedures*, edited by Frank Barnaby, 7–36. Basingstoke, UK: Palgrave Macmillan, 1990.

Ward, Barbara. *Spaceship Earth*. New York: Columbia University Press, 1966.

Wattendorf, Frank L. "Opening Address." in *Polar Atmosphere Symposium*, edited by Sutcliffe, vii–viii.

Weart, Spencer. *The Discovery of Global Warming*. Cambridge, MA: Harvard University Press, 2008.

Weekes, K., ed. *Polar Atmosphere Symposium. Part 2, Ionospheric Section*. New York: Pergamon, 1957.

Weinberger, Hans. "The Neutrality Flagpole: Swedish Neutrality Policy and Technological Alliances, 1945–1970." In *Technologies of Power: Essays in Honor of Thomas Parker Hughes*, edited by Michael Thad Allen and Gabrielle Hecht, 295–332. Cambridge, MA: MIT Press, 2001.

Weir, Gary. *An Ocean in Common: American Naval Officers, Scientists and the Ocean Environment*. College Station: Texas A&M University Press, 2001.

Weisman, Steven R., ed. *Daniel Patrick Moynihan: A Portrait in Letters of an American Visionary*. New York: Public Affairs, 2010.

Werskey, Gary. *The Visible College: A Collective Biography of British Scientists and Socialists of the 1930s*. London: Free Association, 1988.

Weyler, Rex. *Greenpeace: How a Group of Journalists, Ecologists, and Visionaries Changed the World*. Emmaus, PA: Rodale, 2004.

Wheeler, Michael O. "NATO Nuclear Strategy, 1949–1990." In *History of NATO*, edited by Schmidt, 3: 121–40.

Wineland, David. "Norman Ramsey (1915–2011)." *Nature* 480 (2011): 182.

Wittmann, Klaus. "The Road to NATO's New Strategic Concept." In *History of NATO*, edited by Schmidt, 3: 219–37.

Wittner, Lawrence S. *Toward Nuclear Abolition*. Stanford, CA: Stanford University Press.

Wolfe, Audra J. *Competing with the Soviets: Science, Technology, and the State in Cold War America*. Baltimore, MD: Johns Hopkins University Press, 2015.

Wolff, Torben. "The Birth and First Years of the Scientific Committee on Oceanic Research." *SCOR History Report* 1 (2010): 1–96.

Wynne, Brian. "Misunderstood Misunderstanding: Social Identities and Public Uptake of Science." *Public Understanding of Science* 1 (1992): 281–304.

Zelko, Frank. *Make It a Green Peace! The Rise of Countercultural Environmentalism.* New York: Oxford University Press, 2013.

Zierler, David. *The Invention of Ecocide: Agent Orange, Vietnam, and the Scientists Who Changed the Way We Think about the Environment.* Athens: University of Georgia Press, 2011.

Zuckerman, Solly. *Monkeys, Men, and Missiles: An Autobiography, 1946–88.* New York: Norton, 1989.

Zuckerman, Solly. *Scientists and War: The Impact of Science on Military and Civil Affairs.* London: Hamish Hamilton, 1966.

Index

NATO Secretary General, 41, 64, 75–76, 81,
83, 90, 100, 104, 138, 140, 161, 167
NATO Standing Group, 24, 33, 65, 68, 69;
meteorology committee of, 24, 60
NATO strategies: Active Engagement,
Modern Defence, 161; Deep Strike, 141;
enlargement, 160, 163; flexible response,
63–66, 140; massive retaliation, 17, 22,
55, 63. *See also* strategy of tension
NATO Supreme Allied Commander for
the Atlantic (SACLANT), 17, 50, 52;
SACLANT Centre (SACLANTCEN), 50–
51, 53, 117, 164
NATO Supreme Allied Commander for
Europe (SACEUR), 64
Néel, Louis, 7, 41, 69, 72, 74, 80, 97f, 100, 135
Netherlands, 6, 16, 26, 51, 131, 141; Dutch
National Defense Research Organization
(TNO-RVO), 21
Nicholson, Max Edward, 15, 85, 87
Nierenberg, William Aaron, 30, 34, 55, 61,
70, 72, 96, 110, 114, 121, 141, 152–53,
168–69
Nihoul, Jacques C., 112
Nixon, Richard, 78–79, 87–88, 90–91, 93–94,
103–4, 107, 109–10, 141, 152, 167, 204n68
non-proliferation. *See* nuclear non-
proliferation treaties
Norstad, Lauris, 64–65, 196n24
North Atlantic Council (NAC), 7–8, 16–18,
25–26, 28–30, 37, 50, 53, 58, 61, 65, 71,
73, 76, 81–82, 88–90, 105, 114, 116, 126,
163, 198n69
Norway, 6, 13, 16–18, 24–26, 29, 41, 51, 95,
127, 156, 163; Christian Michelsen Institute, 54, 132; Kjeller station, 45, 128; Norwegian Defence Research Establishment
(NDRE), 24, 86; Norwegian Institute of
Meteorology, 60
Novaya Zemlya, 23
nuclear energy. *See* atomic energy
nuclear physics, 27, 41. *See also* nuclear
research
nuclear test ban negotiations, 29
nuclear test facilities, 23
nuclear tests: in Mururoa, 154; in Reggane,
62; *Starfish Prime*, 67; *Tsar Bomba*, 48
nuclear non-proliferation treaties: antiballistic missile treaty, 127; Intermediate-
Range Nuclear Forces Treaty, 156; Non-

Proliferation Treaty, 134; Strategic Arms
Limitation Talks (SALT) 1, 127
nuclear weapons, 5–6, 16–18, 23, 29, 34, 42,
52 64–66; proliferation of, 26, 34, 165. *See
also* NATO Nuclear Planning Group
nuclear winter. *See* climate change

oceanography, 3, 22, 34, 42, 50–53, 56, 59,
65, 95, 96, 110–12, 114, 170; at NATO, 51,
55, 61, 116; in the Soviet Union, 52–53
oil spills, 89–90, 95, 105–6, 112, 132–33, 135,
154, 168. *See also* environmental disasters
Ølgaard, Povl L., 121, 141
operational research, 3, 21, 31–32, 51, 183n38
Oppenheimer, Carl Henry, 97
Optical Atmospheric Quantities in Europe
(OPAQUE), 130–32
Orbiting Radio Beacon Ionospheric Satellite
(ORBIS). *See under* satellites
Organization for Economic Co-operation
and Development (OECD), 81, 109, 144
Organization for Security and Co-operation
in Europe, 163
Özdas, Nimet, 7, 82, 97, 115, 116f, 119, 135,
140–42, 145, 169
ozone, 157; ozone hole, 137; ozone-depleting
substances, 145, 157; ozone sounding, 62

Pâques affair, 73
Partnership for Peace agreement, 161
Patterson, Clair Cameron, 118
Pearson, Lester, 28. *See also* Three Wise Men
Peccei, Augusto, 100, 204n78
Perrin, Francis, 41
Peru, 125
Pestel, Eduard, 79, 80, 93–96, 100–101, 110,
113–14, 169. *See also* NATO Science Com-
mittee: Pestel Working Group
Piloty, Hans, 38
Plass, Gilbert, 131
Poland, 156, 160
Polar Atmosphere symposium. *See* atmosphere
Polaris. *See under* missiles
Pollack, Herman, 81–82
pollution: air, 89, 106, 144, 154, 166; sea,
83, 89–90, 95, 97, 103, 105, 106, 111, 112,
114, 117, 130. *See also* Convention on the
Prevention of Pollution at Sea by Oil;
Long-Range Transboundary Air Pollu-
tion Convention